MOLECULAR BASIS OF INHERITED DISEASE

SECOND EDITION

IN FOCUS

Titles published in the series:

*Antigen-presenting Cells

*Complement

DNA Replication

Enzyme Kinetics

Gene Structure and Transcription 2nd edn

Genetic Engineering

*Immune Recognition

*B Lymphocytes

*Lymphokines

Membrane Structure and Function

Molecular Basis of Inherited Disease 2nd edn

Molecular Genetic Ecology

Protein Biosynthesis

Protein Engineering

Protein Targeting and Secretion

Regulation of Enzyme Activity

*The Thymus

*Published in association with the British Society for Immunology.

Series editors

David Rickwood

Department of Biology, University of Essex, Wivenhoe Park,
Colchester, Essex CO4 3SQ, UK

David Male

Institute of Psychiatry, De Crespigny Park, Denmark Hill,
London SE5 8AF, UK

MOLECULAR BASIS OF INHERITED DISEASE

SECOND EDITION

Kay E. Davies

Molecular Genetics Group, Institute of Molecular Medicine, John Radcliffe Hospital, Oxford OX3 9DU, UK

Andrew P. Read

Department of Medical Genetics, St Mary's Hospital, Hathersage Road, Manchester M13 0JH, UK

IRL PRESS
—at—
OXFORD UNIVERSITY PRESS

Oxford University Press
Walton Street, Oxford OX2 6DP

Oxford is a trade mark of Oxford University Press

Published in the United States
by Oxford University Press, New York

A catalogue record for this book is available from the British Library

Library of Congress Cataloguing in Publication Data
Davies, K.E. (Kay E.)
Molecular basis of inherited disease/Kay E. Davies, Andrew P.
Read. – 2nd ed.
9.
Includes bibliographical references and index.
1. Genetic disorders. 2. Molecular genetics. 3. Chromosome mapping.
I. Read, A.P. (Andrew P.) II. Title.
[*DNLM: 1. Chromosome Mapping. 2. Genetic Markers. 3. Hereditary
Diseases – genetics. QZ 50 D256m*]
RB155.5.D38 1992   616'.042 – dc20   92 – 1453

ISBN 0 19963307 X

Typeset and printed by Information Press Ltd, Oxford, England.

Preface to the second edition

In the three years since the first edition of this book was published, progress in our understanding of inherited disease has been increasing exponentially. Many of these advances are due to the technological improvements in analysis such as the wider applications of PCR, higher resolution of *in situ* hybridization, and the cloning of large DNA fragments in YACs. Many more genetic disorders can now be diagnosed *in utero* with DNA markers, and genes mutated in most of the major genetic diseases have been identified. Mutation mechanisms include point mutations in the globin genes, high frequencies of deletions in the dystrophin gene, unstable elements in the androgen receptor and at the fragile X site, and the duplication of chromosomal material in Charcot–Marie–Tooth disease. This new edition attempts to review the methods of analysis of the human genome in the light of these recent developments. We also describe the use of animal models in the analysis of human disease and in the development of new therapies in the future.

   This is a rapidly expanding and exciting field. We hope that students (clinical and biochemical) will enjoy reading this book, whether they are working directly on human genetic disease or just interested in obtaining sufficient background knowledge in order to follow the literature.

Kay E. Davies
Andrew P. Read

Contents

3. The identification of genes in human inherited disease   37

4. The molecular basis of human inherited disease   63

Abbreviations

A	adenine
α-1AT	α-1 antitrypsin
BMD	Becker muscular dystrophy
bp	base pair
C	cytosine
cDNA	complementary DNA
CF	cystic fibrosis
CGD	chronic granulomatous disease
cM	centiMorgan
CMGT	chromosome-mediated gene transfer
DMD	Duchenne muscular dystrophy
FACS	fluorescence-activated cell sorter
G	guanine
Glu	glutamic acid
HbS	haemoglobin-S
HPFH	hereditary persistence of fetal haemoglobin
HPRT	hypoxanthine guanine phosphoribosyl transferase
HTF	*Hpa*II tiny fragment
IVS	intervening sequence (intron)
kb	kilo base pair
kD	kilo Daltons
LDL	low density lipoprotein
Lys	lysine
PCR	polymerase chain reaction
PERT	phenol-enhanced reassociation technique
PFGE	pulsed-field gel electrophoresis
poly(A)$^+$RNA	polyadenylated mRNA
Pro	proline
RFLP	restriction fragment length polymorphism
T	thymine
Val	valine
VNTR	variable number tandem repeat

1

Genes and markers

1. Forward and reverse genetics

Understanding the molecular basis of inherited disease ultimately requires us to know everything about every gene: its chromosomal location, its DNA sequence, how its expression is controlled, and what its product does. For a given disease, there are two routes to this knowledge, often called 'forward' and 'reverse genetics'. Although different authors have used the term 'reverse genetics' to mean slightly different things, it remains a useful label for a very distinctive strategy in human genetic research (*Figure 1.1*).

The classic human example of 'forward' genetics, the study of the haemoglobinopathies, is described in Section 2 of Chapter 4. Another example is the study of the many inborn errors of metabolism, where the defective enzyme has been identified by standard biochemical methods. For many diseases, however, exhaustive investigations of biochemistry and pathology have failed

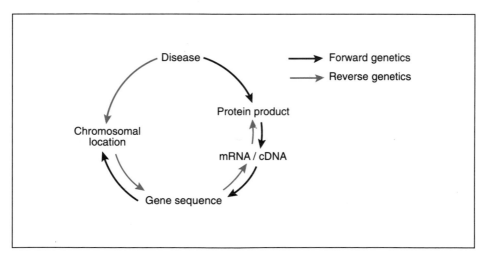

Figure 1.1. Forward and reverse genetics.

1

to reveal the underlying defect. Cystic fibrosis and Duchenne muscular dystrophy are important examples of this, described in detail later (see Chapter 4). Here, the route to success lay through mapping the chromosomal location of the gene, cloning the gene knowing nothing about it except its location (positional cloning), and identifying the product by reference to a table of the genetic code. The product could then be isolated by expressing the cloned gene in an artificial expression system.

Reverse genetics depends crucially on the techniques of genetic mapping and detection of DNA sequence variants, which form the subject of this chapter and Chapter 2.

2. Genetic markers

A genetic marker is any simply-inherited character which is studied not for its intrinsic interest but in order to track the inheritance of a particular segment of a chromosome through a family. A useful marker should be easily studied on readily available material (e.g. a blood sample rather than a brain biopsy); it must be inherited in one of the simple Mendelian patterns described in Section 3 of Chapter 2; and, crucially, it must be polymorphic. The essence of a useful marker is that it should show differences between people, so that the differences can be tracked through a pedigree.

2.1 Why use markers?

Genetic mapping—finding the chromosomal location of genes—is the first stage in reverse genetics. As explained in detail in Section 4 of Chapter 2, genetic mapping depends on studying the way in which pairs of characters, not single characters, segregate in a pedigree. Since one is interested in defective genes, it seems logical to study pairs of defective genes, and this is the normal procedure with experimental organisms like *Drosophila* or mice. However, human genetic diseases are rare, and families in which two different genetic diseases are segregating are exceedingly rare. There simply are not enough of them to make this approach to mapping a practicable proposition, so instead genetic markers are studied. These are natural non-pathological polymorphisms where different forms segregate in the majority of ordinary families. The pair of characters used for mapping can be either a genetic disease and a marker or two markers.

In mapping, much depends on the quality of the pedigrees used. The maximum information comes from three-generation families with all four grandparents available and a large number of children. Although such families can be found, they rarely also have interesting genetic diseases. A general strategy of the human genome project is to perform marker – marker mapping in these 'ideal' families to establish a framework of mapped markers covering every chromosome. The framework can then be used for disease – marker mapping in families with a less ideal structure. Thus, in 1983 Gusella and co-workers demonstrated linkage between Huntington's disease and a marker, a DNA polymorphism revealed by the randomly generated DNA probe, G8 (1). Studies on somatic cell hybrids showed that G8 mapped to the short arm of chromosome 4. A rational search of chromosome 4 could then be initiated in order to discover and characterize the gene involved in Huntington's disease.

2.2 The polymorphism information content of a marker

The usefulness of a marker depends on its informativeness. A marker is informative when it is possible to tell which of his or her two alleles a person has passed on to a child. Consider a marker with alleles (alternative forms) 1, 2, 3, and so on. Each person has two alleles. In *Figure 1.2*, only in pedigrees (d) and (e) is it possible to identify which of his two alleles the father passed on to his daughter. For us to be able to derive useful information, the parent in question must be heterozygous (i.e. have two different alleles, as in pedigrees (c), (d), and (e), but not (a) or (b)), and in addition both parents and the child must not all be the same heterozygote (as in pedigree (c), where it is not possible to say whether allele 1 came from the mother and 2 from the father, or *vice versa*). The polymorphism information content (PIC) measures the probability that this is true. PIC is a number between 0 and 1: the nearer it is to 1, the better the marker. Strictly, PIC is defined as

$$PIC = 1 - \sum_i p_i^2 - \sum_i \sum_{j=i+1}^{n} 2p_i^2 p_j^2$$

where p_i is the gene frequency of the ith allele (2). Intuitively, informativeness is maximized if the marker has many alleles rather than just two, and if each allele is relatively common in the population.

2.3 Types of genetic markers

Blood groups were the first genetic markers to be used. The problem with using them is that there are not many highly polymorphic systems, and antisera for typing the rarer ones can be difficult to find. In the 1960s and 1970s, tissue types and electrophoretic variants ('electromorphs') of serum proteins received much

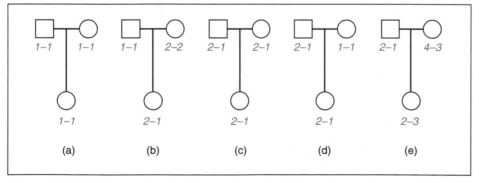

Figure 1.2. Pedigrees showing informative and uninformative marker types (□ males; ○ females).

attention, but nowadays genetic mapping relies almost exclusively on DNA sequence variants. As well as being relatively easy to study by standard methods using any available biological sample, they have the key advantage of numbers: there are so many DNA polymorphisms that one can always be found, given enough work, in any chromosomal location which is under study (2). A second important advantage is that the chromosomal location of a DNA sequence can readily be found by the physical mapping methods described in Section 2 of Chapter 2. Thus, genetic and physical maps can be tied together in a way which was difficult to achieve with protein markers. Ideally, markers should be *sequence-tagged sites*, that is, known DNA sequences located at a cytologically defined chromosomal position and revealing a high-PIC polymorphism.

If corresponding DNA sequences of two unrelated people are examined, it might be predicted that they would be generally identical, but with a constant low probability that any given nucleotide would differ between the two. In fact, this is not the case. About one nucleotide in 200 is polymorphic, with two or more variants common in the population, while all the rest are nearly always the same. This must indicate something about human evolution—but more practically, it is these DNA sequence polymorphisms which make human genetic mapping, and hence reverse genetics, possible.

Two types of DNA polymorphisms are seen (*Figure 1.3*). Most commonly there are two alternative possibilities for a particular nucleotide. Sometimes, however, the polymorphism affects a run of simple, tandemly repeated sequences. Such repeats occur frequently throughout the genome, and in many (but not all) cases the repeat number varies between people. The repeat unit may be a dozen or so nucleotides long (minisatellites) (3,4) or only two or three nucleotides (microsatellites, most commonly runs of $(CA)_n$) (5), or even variable-length runs of a single nucleotide, usually A.

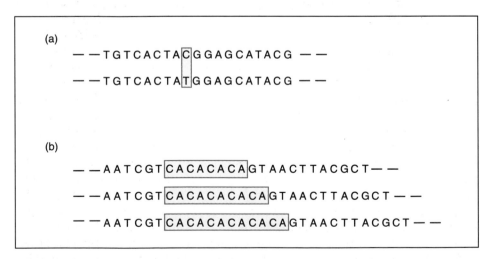

Figure 1.3. Two types of polymorphisms: (a) single base change (b) variable number of tandem repeats.

3. Studying DNA sequence variants

The essential problem in studying human DNA is that there is so much of it—6×10^9 base pairs in a diploid cell, or 3×10^9 base pairs in a single human genome. A means has to be devised to look at the sequence of interest against the background of a vast excess of irrelevant DNA. There are two ways of doing this: either the sequence of interest is picked out by hybridizing it specifically to a labelled probe, or alternatively it is amplified so that it constitutes a significant proportion of the total DNA.

3.1 DNA hybridization assay

About 35 per cent of the human genome is made up of repetitive DNA, sequences present in many copies which may be clustered or scattered across the chromosomes. The remaining 65 per cent is unique sequence DNA, such that any 15-base sequence is likely to occur only once in the genome. Hence, most probes of 15 or more bases hybridize to only one sequence in the genome (two copies per diploid cell). Hybridization is detected by immobilizing denatured DNA on a nylon membrane which is then immersed in a solution of the labelled, single-stranded DNA probe. Most usually the label is radioactive, ^{32}P or sometimes ^{35}S; other labelling systems rely on enzyme activity, fluorescence, or chemiluminescence, often coupled to the probe through a biotin – avidin system.

Very short probes (20 or so nucleotides) require a perfect match to hybridize. Thus, oligonucleotides can be designed to discriminate between two sequences differing by only a single base change. These are called allele-specific oligonucleotides (ASOs) and are made by chemical synthesis. They are technically very difficult to use on natural genomic DNA, but work well if the target sequence is first amplified by PCR (see Section 3.2). Radiolabelled ASOs are widely used as direct tests for particular pathogenic mutations (6). Longer probes will hybridize despite some mismatches; the extent of hybridization depends on the stringency of washing. Typically, a 0.5 – 2 kb stretch of genomic DNA or cDNA, cloned in a plasmid, is used. Radiolabelled copies are made using DNA polymerase, with random hexanucleotide primers, in the presence of ^{32}P labelled ATP. After denaturing by brief boiling, under the correct incubation conditions, such a probe will hybridize to the homologous genomic fragment in DNA from any person, regardless of any small sequence variants present.

The test DNA can simply be spotted onto the nylon membrane (dot blots) but most usually it is first run out on an electrophoretic gel, often after digestion with a restriction endonuclease, and then denatured and transferred to the membrane by Southern blotting (*Figure 1.4*). Sometimes, particularly with ASOs, the aim is to see whether the probe does or does not hybridize (i.e. whether the target sequence is present or absent in the test DNA); more usually the target is always present, but has alternative forms which constitute the alleles of a genetic marker, and the aim is to see which allele is present in the test DNA.

If DNA is cut using a restriction endonuclease, any fragment which contains a tandem repeat with variable copy number (VNTR polymorphism (4)) will show

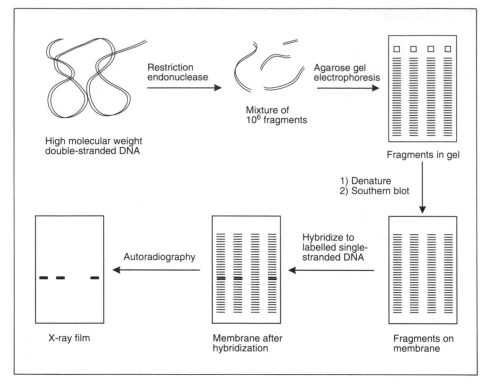

Figure 1.4. Restriction digest, gel electrophoresis, and Southern blot.

size variation between different people. Probes which hybridize to such fragments make excellent genetic markers, with many possible alleles. After gel electrophoresis and Southern blotting, hybridizing bands are seen in different positions arising from different versions of the VNTR sequence (*Figure 1.5a*). Sequences lacking tandem repeats may also appear on differently sized fragments after restriction enzyme digestion, if a polymorphic nucleotide creates or destroys a restriction site within (*Figure 1.5b*) or close to (*Figure 1.5c*) the hybridizing fragment. These are the classic restriction fragment length polymorphisms (RFLPs). There are generally only two alleles, corresponding to presence or absence of the polymorphic restriction site.

3.2 Polymerase chain reaction (PCR) assay

Figure 1.6 shows the principle of PCR (7). It depends on the observation that DNA replication requires a short primer sequence, which the polymerase then extends. To amplify a given sequence, synthetic primers are introduced, typically 20–25 nucleotides long, which match a pair of sequences flanking the target and orientated so that replication in the 5′ to 3′ direction copies the target DNA sequence. PCR proceeds by many rounds of annealing primers, polymerization,

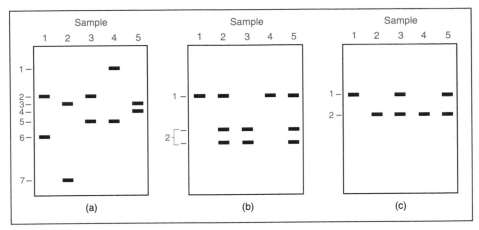

Figure 1.5. Autoradiographic pattern with DNA from five people digested with a restriction enzyme and hybridized to different radiolabelled probes. (a) VNTR polymorphism (b) variable restriction site within the sequence hybridizing to the probe (c) variable restriction site flanking the sequence hybridizing to the probe.

and denaturation. Sequences up to 1 kb long amplify well, longer ones with more difficulty, up to a maximum of about 5 kb. Theoretically, the amount of the target sequence doubles in each round; in practice, 20–30 rounds will amplify the desired target 10^5-fold from 1 µg of genomic DNA, by which time it constitutes the bulk of all the DNA present. It is, of course, necessary to know enough of the sequence of the target to be able to design the correct primers, although it may be possible to amplify completely unknown sequences by *Alu*–PCR (Section 7.4 of Chapter 3).

During the relatively short time since the PCR technique was devised, it has revolutionized many aspects of molecular genetics and spawned innumerable variants. From the standpoint of human disease studies, its main uses are to produce enough of a template to allow single base changes to be sought (Section 3.3) and to provide an alternative and simpler means of scoring RFLPs. To score RFLPs, the sequence containing the polymorphic restriction site is amplified, then the product is digested with the restriction enzyme and run out on a gel to see whether it has been cut. Very little starting material is needed, and the PCR product can be visualized in the gel simply by UV fluorescence in the presence of ethidium bromide, with no need for Southern blotting or probes (*Figure 1.7*). Variable number tandem repeat (VNTR) polymorphisms simply give differently-sized PCR products, and PCR has been used particularly for the CA-repeat microsatellites (5).

3.3 Detecting single base changes in DNA

Base changes which create or abolish restriction sites can be detected easily by Southern blotting or PCR as described previously. However, only perhaps one in six of all nucleotides falls within a palindromic sequence of the type recognized by restriction enzymes. A number of methods have been used to detect other

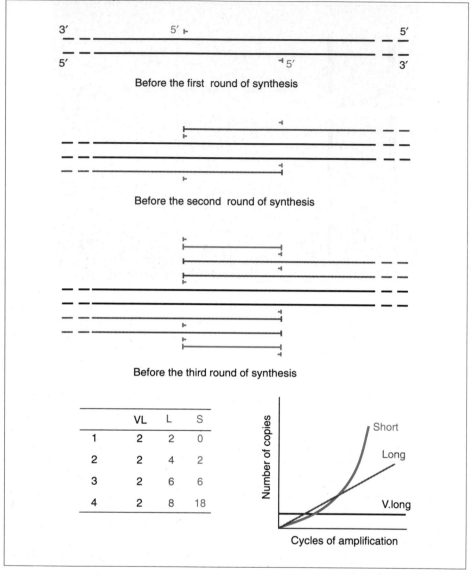

Figure 1.6. The polymerase chain reaction (PCR). Synthesis by DNA polymerase works by extending a short primer (⊢). Long (L) chains are synthesized on the initial very long (VL) template; short (S) chains are synthesized on L or S templates. After several rounds of PCR, the product consists almost entirely of short (S) chains.

single-base changes. They can be divided into methods of checking for a known change, and methods of seeking unknown changes which might occur anywhere within a gene.

The use of allele-specific oligonucleotides for scoring known variants has been

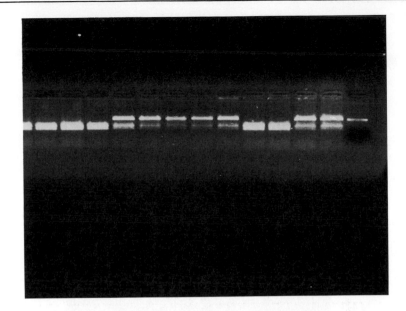

Figure 1.7. PCR-amplified sequence containing a polymorphic restriction site. Exon 11 of the CFTR (cystic fibrosis) gene was amplified and digested with *Hin*dII. Certain cystic fibrosis mutations (G551D and R553X) abolish a normal *Hin*dII site, giving a slower-moving band. Products are seen by ethidium bromide-UV fluorescence.

described (Section 3.1). An alternative is *amplification refractory mutation system* (ARMS) (8) (*Figure 1.8*). ARMS depends on the fact that PCR primers work only if the 3′ end nucleotide is a correct match to the template. Some degree of mismatching is tolerated elsewhere, but the 3′ end must match. Hence a pair of primers can be designed whose 3′ ends match two alternative versions of a sequence, and each will amplify only its correct target. A little juggling of mismatches elsewhere in the sequence is required for good specificity. ARMS provides a non-radioactive method of scoring particular variants that can be automated, and has been suggested as a method for screening the population for carriers of mutations causing cystic fibrosis.

Scanning a gene for unknown mutations requires different techniques. DNA sequencing can always be used, but is laborious with the present technology. Alternatives are single-strand conformation polymorphisms and methods for detecting mismatches in heteroduplexes between the trial sequence and a standard reference sequence.

3.3.1 Single-strand conformation polymorphisms (SSCPs)

Single-stranded DNA tends to fold up, just like a polypeptide, into complex structures stabilized by intramolecular weak bonds. Mostly these will be base-pairing hydrogen bonds, and so the precise structure formed will depend on the nucleotide sequence. If such structures are electrophoresed through a non-

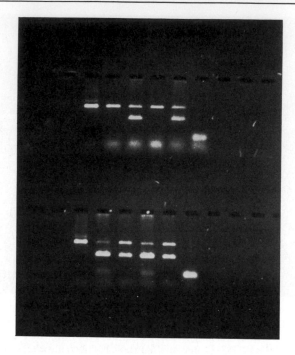

Figure 1.8. ARMS test for the G542X cystic fibrosis mutation. Exon 11 of the CFTR gene was amplified using G542X-specific primers (top) and normal-specific primers (bottom). All reactions include a control amplifiable sequence (slower band) to confirm that the PCR is working. The two outer lanes are blanks to control for contamination. Samples in lanes 3 and 5 are heterozygous for G542X, but none of the other samples contains this mutant sequence.

denaturing gel, their mobility will depend on not just the chain length, as in normal DNA electrophoresis, but also the shape of the molecule, and hence on the sequence. This forms the basis of an elegant method for picking out abnormal sequences (9).

The sequence of interest is amplified by PCR using either a labelled primer or labelled nucleotide. The double-stranded labelled product is denatured with formamide and heat, then run in a non-denaturing polyacrylamide gel. If a series of samples are run together, any that have a sequence change affecting the conformation will show an abnormal mobility. Provided that the temperature of the gel is carefully controlled, most single base changes in PCR products up to 200 bases long show up. The precise abnormality has to be determined by sequencing, but this is a good technique for screening a series of samples for unknown mutations (*Figure 1.9*).

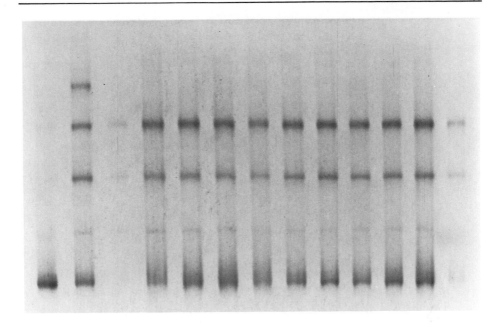

Figure 1.9. Single-strand conformation polymorphism (SSCP) analysis of part of the neurofibromatosis gene. The left hand lane is a control blank. The sample in lane 2 shows an abnormal band; this person turned out to be heterozygous for a new point mutation in the gene. Photo by courtesy of Dr Nalin Thakker.

A variation of this method is to look for abnormal mobility of heteroduplex DNA. Heterozygotes for the common cystic fibrosis mutation, a three-base deletion (see Section 3.5 of Chapter 4) can easily be detected this way. A region containing the mutation is amplified, and the product is denatured and allowed to renature. This allows heteroduplexes to form which have a noticeably lower mobility in agarose gels than either the normal or the mutant homoduplex. This method may be applicable to screening heterozygotes for unknown mutations (10).

3.3.2 Ribonuclease A cleavage of mismatched heteroduplexes

Ribonuclease A will cleave RNA in a DNA–RNA duplex at a point of mispairing (11). *Figure 1.10* shows how this is used. A single-stranded RNA probe is synthesized by *in vitro* run-off transcription of a cloned DNA fragment (with the normal sequence) in an appropriate vector. The probe is hybridized to the test DNA. If the test DNA contains a single base change, there will be a mismatch, which will be cleaved by ribonuclease A. The products are run on a denaturing gel, where the mutation will be revealed because the RNA runs as two bands instead of one. Their sizes pinpoint the position of the mutation.

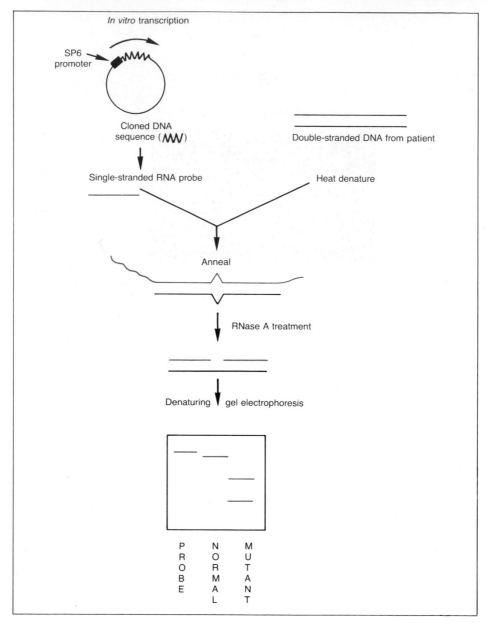

Figure 1.10. Ribonuclease A cleavage.

3.3.3 Chemical cleavage of mismatched heteroduplexes

Mispaired bases in a double helix are more reactive than correctly paired bases. If a reaction is used which leads to cleavage of the sugar – phosphate backbone then, as with the ribonuclease A method described previously, diagnostic

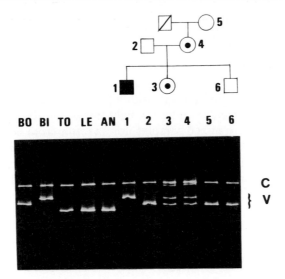

Figure 1.11. Denaturing gradient gel electrophoresis (DGGE) of part of the Factor IX gene in families segregating haemophilia B. The right hand lane shows the normal band positions. Abnormal lower bands are seen in seven tracks, including two doublets from carrier females. Photograph courtesy of Dr Michel Goossens and Academic Press Inc.

fragments are produced. Hydroxylamine reacts specifically with mismatched cytosine in the DNA double helix, and osmium tetroxide reacts with mismatched thymine (and more slowly with mismatched cytosine). In either case, treatment of the reaction product with piperidine breaks the DNA at the position of the mismatch.

To use this method, a heteroduplex is made between the test DNA and a radiolabelled probe, and after reaction the products are run on a denaturing sequencing gel (12). It is convenient to run a sequencing ladder of the probe on the same gel. If the labelled probe contains a mismatched C or T, a cleavage fragment can be seen whose position can be read off the sequencing ladder. If the labelled probe contains a mismatched A or G, then the unlabelled test strand but not the probe will be cleaved. These mismatches could be detected by using a probe complementing the opposite-sense test strand. Thus, in principle, all mismatches might be detectable with two complementary probes.

3.3.4 Denaturing gradient gel electrophoresis

This method has several variants. Homoduplexes or heteroduplexes can be analysed, and the denaturant can be chemical (usually formamide) or high temperature. A good example is described by Attree *et al.* (13). PCR-amplified fragments of the Factor IX gene from members of families with haemophilia B were run on polyacrylamide gels containing a 20–80 per cent gradient of formamide at a constant 60°C. Bands at abnormal positions were seen by ethidium bromide-UV fluorescence in samples from affected males and carrier females in some of the families (*Figure 1.11*). DNA sequencing revealed single-base substitutions.

An alternative version uses a temperature gradient, and this can run either in the direction of electrophoresis (cold to hot) or perpendicular to this. In the latter case, a single wide slot is used, with a large volume of sample; partially melted molecules run slower, so under UV light the sample draws its own melting curve across the gel. With mixed samples, for example from heterozygotes, the curve breaks up into several components, corresponding to each homoduplex and heteroduplex. Again, this method points out the presence of a sequence variant, but DNA sequencing is then necessary to locate and identify it.

4. Further reading

Davies, K.E. (ed) (1988) Genome analysis—a practical approach. IRL Press, Oxford.

5. References

1. Gusella, J.F., Wexler, N.S., Conneally, P.M., *et al.* (1983) *Nature,* **306**, 234.
2. Botstein, D., White, R., Skolnick, M., and Davies, R.W. (1980) *Amer. J. Hum. Genet.,* **32**, 314.
3. Jeffreys, A.J., Wilson, V., and Thein, S.L. (1985) *Nature,* **314**, 67.
4. Nakamura, Y., Leppert, M., O'Connell, P., *et al.* (1987) *Science,* **235**, 1616.
5. Weber, J.L. and May, P.E. (1989) *Amer. J. Hum. Genet.,* **44**, 388.
6. Cutting, G.R., Kasch, L.M., Rosenstein, B.J., Tsui, L.-C., Kazazian, H.H., and Antonorakis, S.E. (1990) *New Engl. J. Med.,* **323**, 1685.
7. Saiki, R.K., Gelfand, D.H., Stoffel, S., *et al.* (1988) *Science,* **239**, 487.
8. Newton, C.R., Graham, A., Heptinstall, L.E., *et al.* (1989) *Nucl. Acids Res.,* **17**, 2503.
9. Orita, M., Suzuki, Y., Sekiya, T., and Hayashi, K. (1989) *Genomics,* **5**, 874.
10. Keen, J., Lester, D., Inglehearn, C., Curtis, A., and Bhattacharya, S. (1991) *Trends Genet.,* **7(1)**, 5.
11. Myers, R.M., Larin, Z., and Maniatis, T. (1985) *Science,* **230**, 1242.
12. Cotton, R.G.H., Rodrigues, N.R., and Campbell, R.D. (1988) *Proc. Natl. Acad. Sci. USA,* **85**, 4397.
13. Attree, O., Vidaud, D., Vidaud, M., Amselem, S., Lavergne, J.-M., and Goossens, M. (1989) *Genomics,* **4**, 266.

2

Locating genes

1. Introduction: genetic and physical mapping

As described in the previous chapter, mapping a disease locus is a necessary first step in the 'reverse genetic' approach to isolating genes for diseases where no protein product is known. Genetic mapping proceeds by linkage analysis. Pairs or groups of loci are studied in pedigrees; from the results linkage groups are defined, and within each group the order and distance apart of the loci are calculated. Genetic distances are expressed in *centimorgans* (cM): loci 1 cM apart show one per cent recombination. Averaged over the whole human genome, one centimorgan corresponds to one megabase of DNA, but there are at least five-fold variations between different chromosomal regions. Ultimately, the human genetic map should show the location of every gene and marker on the 24 different chromosomes (numbers 1 – 22, X, and Y). The present map is far from complete, and adding to it is a main preoccupation of human genetic research. *Figure 2.1* shows a partial map of one of the best mapped chromosomes, the X chromosome.

Unlike genetic mapping, physical methods map loci in cytogenetic terms (e.g. to Xp21 or 7q22) or DNA units (kilobases, megabases, and so on). The nomenclature of chromosome regions (1) is based on the bands and sub-bands seen by cytogeneticists on suitably stained chromosomes. These are indicated in *Figure 2.1*. Xp21 means sub-band 1 of band 2 of the short arm (p = short arm, q = long arm), of the X chromosome; for more precise work, Xp21 is divided into Xp21.1, Xp21.2, and Xp21.3. However, even the smallest cytogenetically observable band contains several megabases of DNA.

2. Gene mapping by physical methods

Physical methods can be used to map a cloned DNA sequence to a defined point on a chromosome. It is also possible to map an enzyme or an antigen, and

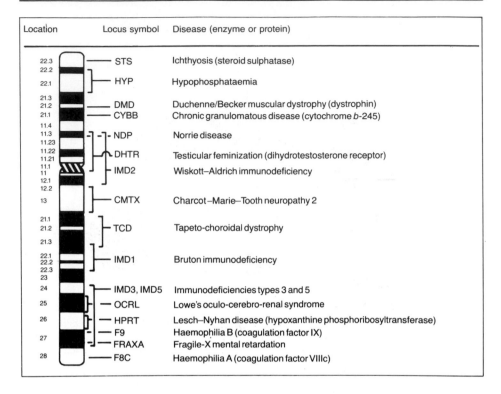

Location	Locus symbol	Disease (enzyme or protein)
22.3 22.2	STS	Ichthyosis (steroid sulphatase)
22.1	HYP	Hypophosphataemia
21.3 21.2	DMD	Duchenne/Becker muscular dystrophy (dystrophin)
21.1	CYBB	Chronic granulomatous disease (cytochrome *b*-245)
11.4 11.3 11.23	NDP	Norrie disease
11.22 11.21	DHTR	Testicular feminization (dihydrotestosterone receptor)
11.1 11 12.1 12.2	IMD2	Wiskott–Aldrich immunodeficiency
13	CMTX	Charcot–Marie–Tooth neuropathy 2
21.1 21.2 21.3	TCD	Tapeto-choroidal dystrophy
22.1 22.2 22.3 23	IMD1	Bruton immunodeficiency
24	IMD3, IMD5	Immunodeficiencies types 3 and 5
25	OCRL	Lowe's oculo-cerebro-renal syndrome
26	HPRT	Lesch–Nyhan disease (hypoxanthine phosphoribosyltransferase)
27	F9	Haemophilia B (coagulation factor IX)
	FRAXA	Fragile-X mental retardation
28	F8C	Haemophilia A (coagulation factor VIIIc)

Figure 2.1. Partial genetic map of the X-chromosome. The complete map would include 150 diseases and 500 cloned sequences. Note the method of numbering the bands.

occasionally a disease phenotype can be physically mapped using a chromosome abnormality.

2.1 In situ *hybridization*

This is the most direct method of mapping. A cloned DNA fragment is hybridized directly to a spread of metaphase chromosomes, and the map location worked out by examining the result under the microscope (2).

The starting point is a standard chromosome preparation. Cells are grown in culture, treated with colchicine to accumulate cells in the metaphase stage of mitosis, fixed and spread on to a microscope slide. The chromosomes are treated with trypsin and stained with Giemsa stain which gives a pattern of dark and light G-bands, from which each individual chromosome can be recognized. Chromosomes are then denatured on the slide to leave the DNA single stranded without destroying the morphology of the chromosomes.

The slide is then incubated with a labelled, single-stranded DNA probe, which hybridizes to any matching sequences in the denatured chromosomes. Fluorescent labelling has largely superseded radiolabelling, and fluorescent *in situ* hybridization (FISH) is one of the most powerful and elegant techniques available

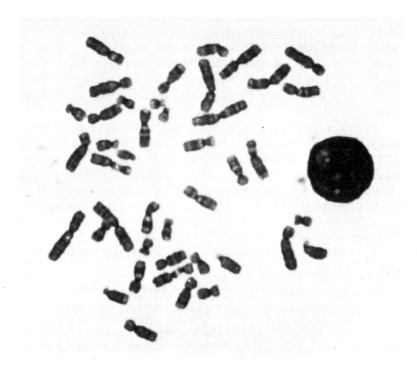

Figure 2.2. Fluorescent *in situ* hybridization. In this simulated photograph, the fluorescent spots are located over the terminal region of the long arm of chromosome 12. Note the two fluorescent spots over the interphase nucleus.

for physical mapping (3). The probe is usually coupled covalently to biotin, and the signal developed using an avidin-carrying fluorescent reporter molecule. One or more layers of antibodies may be included, to amplify the signal. In marked contrast to the older radioactive technology, FISH gives quick and usually unambiguous results which can be scored by eye under a fluorescence microscope (*Figure 2.2*). With more sophisticated equipment, two differently coloured fluorescent probes can be used simultaneously, allowing direct visualization of the order of the probes on the chromosome.

On metaphase spreads, the resolution is 10–20 million base pairs. In interphase nuclei, fluorescent dots can be seen by FISH. These cannot reveal the chromosomal location, but if sets of probes with different fluorescent labels are used, their distance apart and their order can be established with a resolution as precise as 40 kb (4).

2.2 Somatic cell hybrids

Until recently, this has been much the most widely used physical mapping method in human genetics, being much more robust than the earlier radioactive *in situ* methods (5).

If rodent and human cells in culture are fused by treatment with polyethylene glycol, the resulting heterokaryons are unstable and tend to lose only the human chromosomes, but in a more or less random way. Eventually, stable cell lines are produced which contain a full set of rodent chromosomes plus a few human chromosomes. The human chromosomes present must be characterized; this is done under the microscope, and by PCR using sets of primers specific for each human chromosome (6).

Once a collection of well-characterized hybrids has been prepared, cloned human DNA sequences can be tested for hybridization against the bulk DNA of each cell line. Alternatively, the presence of human gene products can be correlated with the presence of particular human chromosomes. *Figure 2.3* shows an example of this method. Of course, for this technique to work, the human DNA must not cross-hybridize to rodent DNA, or the human product must be

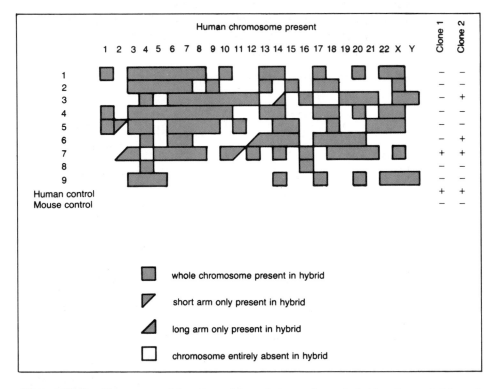

Figure 2.3. Chromosomal location of two clones using a hybrid cell panel. Clone 1 maps to the long arm of chromosome 2; clone 2 maps to chromosome 19 or the long arm of chromosome 12.

distinguishable from any homologous rodent product. If the original human cells contained a translocated or deleted chromosome, the hybrids can be used to map a DNA fragment relative to the chromosomal breakpoint. For very fine mapping, irradiation hybrids can be prepared which contain only small fragments of human chromosomes (see Section 7.3 of Chapter 3).

2.3 Chromosome abnormalities and loss of heterozygosity

Most chromosomal abnormalities affect many genes and produce multi-system syndromes which are not helpful for genetic analysis. Sometimes, however, a specific disease is regularly associated with a very small chromosome abnormality. Prader–Willi syndrome (see Section 3.7 of Chapter 4) is one example. The classic case is retinoblastoma (7), a tumour of the retina, which is often sporadic and apparently non-genetic but which in some families shows autosomal dominant inheritance. In some cases (familial or sporadic) there is a visible deletion of chromosomal band 13q14. Even where no chromosomal abnormality is visible, retinoblastoma tumour tissue is often homozygous for chromosome 13 markers for which the normal tissue of the patient is heterozygous.

The tumour arises from cells which have lost both copies of a tumour suppressor gene at 13q14. Sporadic cases are due to previously normal cells suffering two independent mutations, while in familial cases one mutation is inherited, so that only one further mutation is needed in any one of the genetically predisposed retinal cells to cause the disease. Familial cases arise by one or other of the mechanisms shown in *Figure 2.4.*

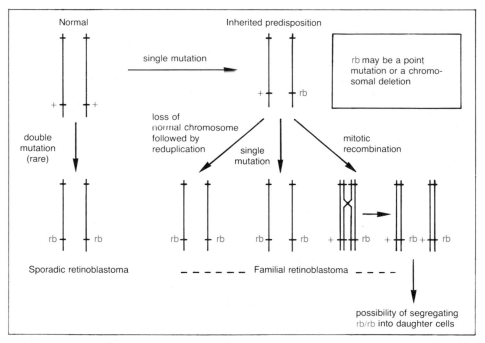

Figure 2.4. Two-hit mechanism in retinoblastoma.

Similar mechanisms are believed to cause many other cancers, and a major research interest at present is to screen tumour tissues for a loss of heterozygosity compared to normal tissues of the patient. By this means, genes involved in both familial and sporadic neoplasias of many types are being mapped and cloned (8).

Chromosome breaks, as well as chromosome deletions, can cause disease. When chromosomes are involved in translocations or inversions the breakpoint may disrupt a gene, or move it into an environment where its expression is inappropriate. For example, the gene for Waardenburg syndrome was mapped to chromosome 2 following the observation of a child who had this dominant disease as an apparent new mutation with no family history, and also had a *de novo* inversion of chromosome 2 (9). Such anecdotal observations need to be confirmed by linkage analysis, because they often turn out to be purely coincidental.

2.4 X-inactivation and X-autosome translocations

In females, one of the two X-chromosomes is genetically inactive ('lyonized') in somatic cells (10). This is a mechanism of dosage compensation, ensuring that XY males and XX females are both normal (remember by comparison what happens if a person has an extra copy of chromosome 21). In the early female embryo, one X-chromosome in each cell is inactivated, and this same chromosome remains inactive in all the cells derived from that particular cell. On the inactive X-chromosome, all except a few genes are transcriptionally silent. The mechanism is at least in part due to a heritable pattern of CpG methylation, and the gene which appears to be responsible for X-inactivation has recently been cloned (11).

X-chromosome inactivation in females is usually random, but X-autosome translocations are a special case. Female carriers of X-autosome translocations preferentially inactivate the normal X-chromosome. Cells which inactivate the translocated X-chromosome suffer chromosomal imbalance because the detached part of this chromosome is not inactivated, and these cells do not survive. If the translocation breakpoint disrupts a gene, then since the other copy of the gene on the intact X-chromosome is always inactivated, the female carrier is as severely affected as a male with the disease. These rare females are of great interest, because the breakpoint shows the location of the disease gene: this is how the Duchenne muscular dystrophy gene was mapped to Xp21 (see Section 3.3 of Chapter 4).

3. Modes of inheritance

3.1 Introduction

Interpreting human pedigree patterns is not straightforward. First, since parents give their children both their genes and their environment, the fact that a character tends to run in families does not prove that it is genetic—it might, for example, be due to a bad diet. Secondly, most characters depend on inter-actions between one or more gene loci, conferring susceptibility, and an environmental trigger. Such characters are exceedingly refractory to current methods of genetic analysis; despite their great clinical importance, very little progress has been made in the genetic analysis of schizophrenia, depressive psychosis, or many other major diseases. Only those which follow the simple (Mendelian) patterns of inheritance have proved amenable to genetic analysis. Here progress has been spectacular, but even Mendelian diseases can give ambiguous pedigree patterns.

It is important to realize that, given enough coincidences (mutations, affected people happening to marry carriers, and so on), any of the patterns in *Figures 2.5* to *2.9* could be seen with any mode of inheritance. Determining the mode of inheritance requires a large amount of data and the statistical technique of segregation analysis (12). Mendelian diseases, though numerous, are individually rare, and with many of them the mode of inheritance is not certain. Insights from molecular biology are proving invaluable for sorting out complex groups of overlapping phenotypes, of which the collagen abnormalities (13) are the cardinal example. For the student, the first rule is to look up the phenotype in McKusick's *Mendelian Inheritance in Man* (14), which is a computerized catalogue of proven and probable Mendelian phenotypes. For each phenotype, there is a catalogue number, which is widely used in the genetic literature, and a few leading references.

3.2 Autosomal codominant inheritance

This is the simplest type of Mendelian pattern. The character is determined by a single genetic locus located on one of the autosomes (i.e. any of the chromosomes except the X or Y sex chromosomes). There may be just two alleles, or a larger number, but the mark of codominant inheritance is that the heterozygote shows both. All good genetic markers are codominant, as are some pathological variants such as haemoglobin-S (HbS); heterozygotes for HbS have sickling trait and homozygotes have sickle cell disease. Remember that dominance and recessiveness are properties of observed characters, not of genes. People often talk loosely of dominant or recessive genes, but this is to be discouraged. Sickle cell anaemia is a recessive disease (it is seen only in HbS homozygotes and not in heterozygotes), but sickling trait is dominant, because it is apparent in the heterozygote.

3.3 Autosomal dominant inheritance

A condition which consistently shows the pedigree pattern in *Figure 2.5* is autosomal dominant. Heterozygotes (*Aa*) manifest the condition. About 1500 autosomal dominant conditions are known (14). Among the examples mentioned in this book are Huntington's disease, achondroplasia, retinoblastoma, and familial hypercholesterolaemia. People affected by rare dominant diseases are almost always heterozygotes. Homozygous mutants (*AA*) may be more severely affected (as with achondroplasia) or indistinguishable from heterozygotes (as with Huntington's disease). In most rare diseases, homozygotes have never been described, either because the necessary marriage of two heterozygous affected

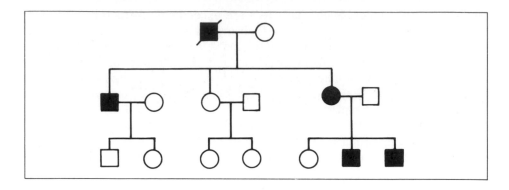

Figure 2.5. Autosomal dominant pedigree pattern. □ male; ○ female; ■, ● affected; ◣ dead; ══ consanguineous marriage; ⊙, carrier (optional).

people has not occurred, or because the homozygotes are so badly affected that they die *in utero*.

The characteristic features of the pedigree are:

(a) a vertical pattern, that is affected patients have an affected parent;

(b) if an affected person marries an unaffected person, the risk of any particular child being affected is one in two;

(c) both sexes are equally affected

(d) both sexes are equally likely to pass on the condition.

If all autosomal dominant pedigrees were as clear as *Figure 2.5*, diagnosis and genetic counselling would be straightforward. Reality is not so simple, however, and pedigree interpretation is often complicated by variable expression, non-penetrance, or mutation. Variable expression means that the condition can be severe in one person and mild in another, perhaps so mild that the person is unaware they have it, yet their children might be severely affected. It is normal for dominant diseases to show variable expression. Part of the skill of the clinical geneticist is to know how variable a particular condition is, and what minimal clinical signs will give a correct diagnosis. Non-penetrance is the extreme of variable expression: although a person has no signs of the condition, they must carry the disease gene because they have an affected parent and an affected child. *Figure 2.6* shows an example of this.

It is not surprising that dominant conditions are variable. To produce a perfect dominant pedigree, a gene must be both necessary and sufficient for the disease. Anyone who has the gene must always show exactly the standard disease phenotype, regardless as to how they live or what genes they carry at all other loci. With many conditions, genetic background or the environment can to some extent modify the effect of the major gene, giving variable expression or occasional non-penetrance. As the effect of these other factors increases, the pedigree becomes less regular, until it becomes more convenient to describe the

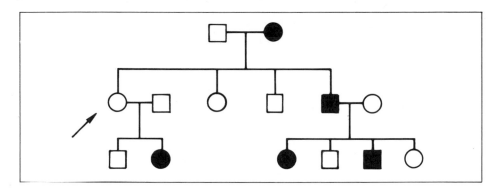

Figure 2.6. Autosomal dominant pedigree with non-penetrance (✗). For definitions of the symbols, see *Figure 2.5.*

condition as 'multifactorial' and non-Mendelian. There is no defined borderline between these categories.

3.4 Autosomal recessive inheritance

Autosomal recessive conditions are determined by a single autosomal locus, but the condition is manifest only in people who are homozygous for the abnormal allele (*aa*). The parents of affected children are usually phenotypically normal carriers (*Aa*), and each child has a one in four risk of being affected.

Over 1000 autosomal recessive conditions are known (14). Examples mentioned in this book include α and β thalassaemia and cystic fibrosis. A typical autosomal recessive pedigree would show a single affected child, with unaffected non-consanguineous parents and no previous family history of the disease. Unless the disease can be recognized clinically or biochemically as a known recessive, there would be no way of knowing that it was genetic at all, until the parents had a second affected child. There must be many unrecognized recessive diseases among the patients seen at genetic clinics. *Figure 2.7* shows a pedigree where the consanguinity and affected sibs (brothers or sisters) make the mode of inheritance more obvious. The distinctive features are:

(a) a horizontal pattern, that is normal parents have one or more affected children;

(b) if an affected person marries, unless they marry a carrier, the children are all unaffected;

(c) an association with consanguineous marriages;

(d) both sexes are equally affected.

When the biochemical basis of a recessive disease is well understood, as with haemoglobinopathies and Tay – Sachs disease, carriers may be identifiable by biochemical testing.

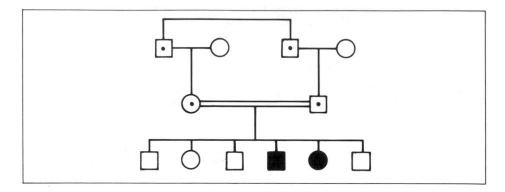

Figure 2.7. Autosomal recessive pedigree pattern. For definitions of the symbols, see *Figure 2.5.*

3.5 X-linked inheritance

Recessive conditions governed by loci on the X-chromosome show a very characteristic pedigree (*Figure 2.8*). Because the pattern is so recognizable, over 200 X-linked recessive conditions have been described (14). Examples used in this book include Duchenne muscular dystrophy, haemophilia A, and chronic granulomatous disease.

The specific features of X-linked recessive pedigrees are:

(a) the disease affects mainly males;

(b) a 'knight's move' pattern: affected males have unaffected maternal uncles;

(c) the disease is transmitted by carrier women who are usually asymptomatic, half the sons of a carrier are affected, and half the daughters are carriers;

(d) if an affected male has children, none of his sons is affected but all his daughters are carriers;

(e) affected women can be born if an affected man marries a carrier woman.

Because of X-inactivation (see Section 2.4), female carriers of an X-linked disease have some cells in which only the normal X is active and some in which only the disease-carrying X is active. As a result, they often show some symptoms. Inactivation is usually random. Sometimes, by bad luck, a carrier will inactivate the normal X-chromosome in most cells, and so suffer symptoms of a disease which is normally recessive ('manifesting heterozygotes').

X-linked dominant inheritance is seen much less often, partly no doubt because the pedigree pattern is nowhere near as distinctive as the recessive pattern. Pedigrees (*Figure 2.9*) resemble autosomal dominant pedigrees except that no sons but all daughters of affected males are affected. The few known examples include the Xg blood group and X-linked hypophosphataemic rickets, together with a few rare diseases (for example, incontinentia pigmenti, focal dermal hypoplasia, and Rett syndrome) which affect only females and are believed to be X-linked dominants, lethal to males.

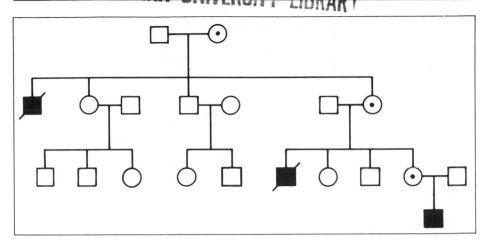

Figure 2.8. X-linked recessive pedigree pattern. For definitions of the symbols, see *Figure 2.5*.

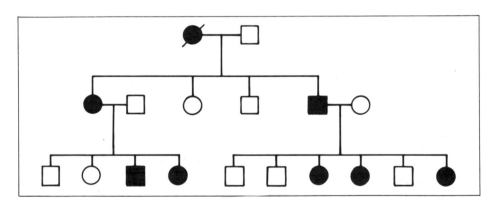

Figure 2.9. X-linked dominant pedigree pattern. For definitions of the symbols, see *Figure 2.5*.

A disease is not necessarily X-linked just because it affects only males—consider cancer of the prostate. Other diseases may have some more subtle reason for affecting only males. If a pedigree shows male to male transmission, then the condition is not X-linked, because a father does not pass his X-chromosome to his sons.

4. Gene mapping by linkage studies

4.1 Introduction
Two genetic loci are said to be linked if they segregate together in pedigrees more often than by random chance. This happens because they lie close together

on the same chromosome. In principle, linkage analysis in man is exactly the same as in *Drosophila* or any other diploid organism, and if the reader is unclear about the principles, then read an account in a standard genetics textbook (see Further Reading). The aim is always to identify and count recombinants in suitable crosses. *Figure 2.10* shows how this might be done in a human pedigree.

The pedigree shows restriction fragment length polymorphism (RFLP) types in a family with an autosomal dominant disease. I_1 has both affected and unaffected children, so she must be heterozygous for the disease gene. She is also heterozygous for the RFLP, thus she is doubly heterozygous. Either two (II_5 and II_8) of the children are recombinants, or else these two are non-recombinants and the other five are recombinants. It is not clear which is true, because I_1 is phase unknown—that is, it is not known whether she carries RFLP allele 1 or allele 2 on the disease-bearing chromosome. The situation is different for the children of II_2. It is known that II_2 received RFLP allele 1 from her mother along with the disease. She is 'phase-known', and it is certain that two of her six children are recombinants. Overall, the pedigree suggests, but does not prove, loose linkage.

The problems in man are, of course, that crosses cannot be set up to suit the investigator, and families are very small. Human linkage analysis relies heavily on exceptional families and on sophisicated statistical methods for combining data from many families.

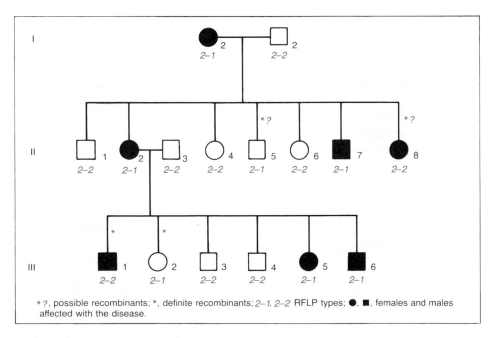

*?, possible recombinants; *, definite recombinants; 2–1, 2–2 RFLP types; ●, ■, females and males affected with the disease.

Figure 2.10. Segregation of an autosomal dominant disease and a possibly linked marker. For definitions of the symbols, see *Figure 2.5*.

4.2 Analysis of linkage studies

The result of a linkage study is typically a table showing the 'lod scores' for a pair of loci at a range of recombination values. Lods are logarithms to the base 10 of odds. The odds in question are the odds that the two loci are linked with this recombination fraction rather than unlinked. Lod scores are usually calculated for recombination fractions 0.0, 0.05, 0.10, up to 0.50 (independent assortment).

This is one branch of genetics which is almost completely dependent on computers. The theory of the lod score method was worked out by Morton in 1955 (15). However, except for the simplest pedigrees, it is really not possible to calculate accurate lod scores by hand and almost all lod score calculations are made using one of two computer programs, LIPED (16) or MLINK (17). These are widely available on mainframes and microcomputers, and should be used by anyone wishing to do linkage analysis in man. Both programs work on the same principle: they calculate the overall likelihood of the data on two alternative assumptions, either that the two loci are linked with the given recombination fraction or that they are unlinked. The ratio of these two likelihoods is the odds that the loci are in fact linked. The logarithm to the base 10 of this ratio is the lod score.

A lod of zero means the assumptions of linkage or no linkage are equally likely ($\log_{10}(1) = 0$). A positive lod score favours linkage and a negative lod is evidence against linkage at the given recombination fraction. The thresholds of significance are $+3$ and -2. At first sight, demanding a lod of $+3$ seems excessively stringent since this means the odds must be 1000:1 in favour of linkage before the evidence is accepted. In fact, Morton chose this threshold with good reason: with a lod of 3.0, the overall probability of linkage is 95 per cent. The reason lies in the low *prior probability* that two loci are linked. In typical human datasets, there is only about a 1 in 50 chance that two randomly chosen loci will happen both to lie close enough on the same chromosome to show linkage. So, somebody claiming linkage has to produce very strong evidence. This can be formalized as Bayesian statistics (18), but it is also simple common sense: if someone tells you he has seen a little green man you will require much more convincing than if he tells you he has seen a little green van.

A plot of the lod score against recombination fraction will take one of the general forms shown in *Figure 2.11*. Curve 1 never crosses the $+3$ or -2 thresholds, and is inconclusive (the remedy is more data). Curve 2 shows data favouring tight linkage; the peak at zero means that no recombinants were seen. Curve 3 excludes linkage closer than 10 cM (recombination fraction 0.1, where the curve crosses the -2 threshold). Curve 4 favours linkage; the peak lod score (symbolized $\hat{z}$) is 5.25 and the most likely recombination fraction ($\hat{\theta}$) is 0.08.

One of the great advantages of lod scores is that they can be added together. The likelihood of two independent sets of data is the likelihood of the first multiplied by the likelihood of the second, and multiplying likelihoods corresponds to adding logarithms. Linkage analysis almost always needs a collection of families. The lod scores from individual families are added together within a

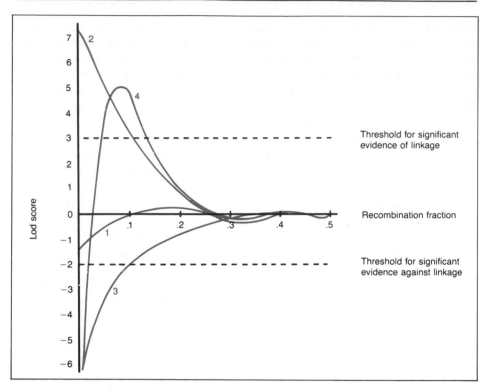

Figure 2.11. Result of linkage analysis. A plot of lod score against recombination fraction for four pairs of loci.

study to get the overall lod score, and when someone else studies more families the new lod scores can be simply added in. This is a great help in linkage work with rare diseases, where in any one country only a few families can be studied.

It should be remembered that when the lod is 3.0 there is still a 5 per cent chance that the loci are in fact not linked. If a set of families with an interesting disease is typed for 20 randomly chosen markers, it is likely that one lod score will be 3.0 purely coincidentally. It is important to preserve a degree of scepticism about lod scores in the 3–5 range, and a great deal of scepticism about claims for linkage based on lod scores below 3. One exception to this is linkage studies of X-linked diseases and markers: the prior probability is high (at least 1 in 10) that two loci will show linkage if they are already known both to be on the X-chromosome, and here a lod score of 2.5 would be satisfactory evidence for linkage.

4.3 Multi-locus linkage analysis

A common type of investigation takes a set of families with a particular disease, and types them for a series of markers, all from one candidate region of a chromosome. Lod scores can be calculated for each marker, but this is not the most efficient use of the data for two reasons. First, particular individuals may be informative (heterozygous) with marker A but not with the closely linked marker B, while others in the same pedigree may be homozygous for A but informative for B. Standard lod score analysis cannot combine information from more than two loci. Secondly, two-point lods are not efficient for deciding the order of a series of linked loci. The recombination fractions at the peak lod are subject to large errors in normally sized datasets, and it is not safe to order loci simply using this parameter. Locus order is important, not only for general mapping but also because once a disease gene has been localized approximately, it becomes important to define the closest flanking markers so that they can be used as starting points in attempts to clone the gene.

Drosophila geneticists use three-point crosses to order loci. In a cross of genotypes *ABC/abc* and *abc/abc*, the rarest class of progeny is the class which requires a double recombination. Suppose this class is found to be *ABc/abc*, then the order of loci must be A – C – B. A variant of the MLINK program described previously makes it possible to run this sort of analysis on human linkage data (19). The program takes an unknown locus (e.g. a disease) and a fixed framework of marker loci, and calculates the overall likelihood of the data for a series of possible positions of the unknown locus. Here, the lod score measures the ratio of the likelihood of the data with the unknown locus at a particular position compared to the likelihood with the unknown locus far away from any of the markers. A plot of lod score against map position (*Figure 2.12*) gives a curve with peaks and deep troughs. The peaks are evidence of linkage (with no linkage the score would remain negative or zero), and the highest peak marks the most likely location for the unknown locus.

Multi-locus lod score analysis is a very powerful technique, particularly now that multiple linked markers which can saturate a chromosome are available. Curiously, perhaps its greatest use is in disproving linkage. Linkage studies necessarily generate large amounts of negative data, and when enough negative data have accumulated, it may become clear that there are relatively few places where the gene can be; this is called exclusion mapping (9,20). Multi-locus linkage analysis is the best way of excluding a locus from a chromosomal region with a framework of markers.

Despite its great power, multi-locus linkage analysis has several drawbacks. One is that the positions of the markers must be known with certainty, which is often not the case. It is also not easy to combine data from several studies, as with two-point lod scores. It can be done using the raw data, but not from the published curves. In general, not too much credence should be placed on the precise genetic distances calculated by multi-locus linkage (or even two-point linkage unless the peak lod score is over 20), but locus orders are much more reliable, at least when the relative odds of two orders come out at 1000:1 or better.

4.4 Sib pair analysis

Sib pair analysis is a simplified form of linkage analysis used for mapping recessive characters, and as a quick test of whether a candidate cloned sequence might be the gene for a disease. The principle is very simple: if one parent has

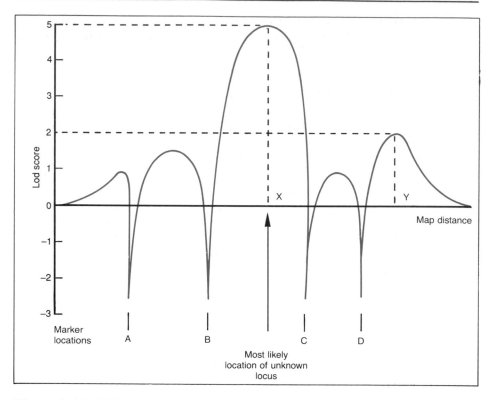

Figure 2.12. Multilocus linkage analysis of pedigrees in which an unmapped disease and four well-mapped markers are segregating. The odds that the disease maps to position X rather than Y is the antilogarithm of the difference in lod scores, that is approximately 1000:1.

marker alleles *ab* and the other *cd*, the children may be *ac*, *ad*, *bc*, or *bd*. There is a one in four chance that two children will have the same types. However, if the two children both have the same recessive condition, they will necessarily inherit the same alleles of markers located close to the gene responsible for the disease. So, sib pair analysis takes pairs of affected sibs and looks for markers which they share more often than one time in four. This was how the cystic fibrosis locus was mapped to chromosome 7. The same method can be used to rule out a candidate gene for a disease. If two affected sibs do not share the same parental copies of the gene, then it cannot be the cause of the disease.

A great advantage of sib pair analysis over conventional lod score analysis is that it does not require the mode of inheritance, gene frequencies, and penetrance to be known. Lod score analysis tests a precise genetic hypothesis, and if the parameters used in the model are wrong then the results are wrong to a greater or lesser degree. This is a particular problem in attempts to apply linkage methods to diseases where the mode of inheritance is not clear cut, as with schizophrenia. Recently, generalizations of sib pair analysis have been developed which can be applied to pedigrees with any structure (21). These may well be the way forward in analysis of the many diseases where there is genetic susceptibility but not clear Mendelian inheritance.

5. Disease associations and linkage disequilibrium

An alternative to linkage for locating disease genes is to study associations. Certain diseases occur more frequently in people who have particular genotypes for markers, that is, there is a statistical association in the population between the disease and a certain marker allele. A classic example is the strong association between insulin-dependent diabetes and the DR3 and DR4 alleles of the HLA – DR locus. More recently, associations have been observed between cleft palate and one allele at the TGFA (transforming growth factor alpha) locus (22).

Apart from pure coincidence, such associations could have three causes. First, the population studied might actually consist of two groups, one of which happened to have a high frequency of both attributes just as part of the general genetic differences between populations; an example would be the association which appears to exist in North America between cystic fibrosis and pale skin. A second possibility is that one attribute is part of the cause of the other (this is the suggestion, so far unproven, for cleft palate and TGFA). Finally, close linkage can produce population associations.

Normally, linkage results in an association within a family but not in the population because of recombination and recurrent mutation. When a disease mutation occurs, it will happen in one particular individual who has some particular collection of individual genetic attributes (say, female, blood group A, HLA – A 2,23, and so on). Supposing the mutation affected a gene on one copy of chromosome 6 close to the HLA – A23 gene. After it has been transmitted through a few generations, the people carrying it may well be male and blood group O, but they are all likely to have HLA – A23.

This is an example of *linkage disequilibrium*. Whether or not a population association is seen depends on whether most of the disease genes in the population are descended from one common ancestral mutation, and, if so, on whether enough time has elapsed for the association to decay. Eventually recombination will break up the association with A23: after n generations only a proportion $(1-r)^n$ of the chromosomes will retain the ancestral A23, where r is the probability of recombination. If r is small enough, disequilibrium can persist over many generations.

As a method of finding disease genes, linkage disequilibrium is very limited. The chance of finding true linkage through population associations is much lower than through conventional linkage studies. This is because conventional linkage analysis effectively looks for associations within families, whose members are probably separated by 2 – 10 meioses from a common ancestor. Therefore, relatively large chromosomal segments are conserved within families, unbroken by recombination and offering targets for linkage. In a population of about 50×10^6, probably 20 – 50 meioses separate supposedly unrelated people through their last common ancestor, so the chromosomal segments which have been transmitted intact are correspondingly smaller. Thus, the chance of a true positive in population association studies is small, while several factors give a large chance of irrelevant or spurious associations. Linkage disequilibrium is, however, often seen in the closing stages of positional cloning, when the markers are getting very close to the disease gene. In these circumstances, it can provide some reassurance that the correct chromosomal region is being investigated, and it may also have a role in risk estimation for genetic counselling (23).

6. Genetic diagnosis

Often it is desirable to know whether somebody carries a disease gene. This may be a carrier test for a recessive disease, a predictive test for a late-onset disease such as Huntington's disease, or a prenatal test on a fetus at risk. There are two essentially different methods of using DNA technology to do this, direct tests and gene tracking.

6.1 Direct tests for mutant alleles

Direct tests are possible only if the gene to be tested has been cloned and characterized. They can detect gross structural changes, particularly gene deletions, and a proportion of subtler changes such as point mutations.

Among the many mutational changes which can render a gene inoperative, physical deletion is the most extreme. Large deletions cause almost all forms of α-thalassaemia (see Section 3.1 of Chapter 4) and 70 per cent of Duchenne or Becker muscular dystrophy (see Section 3.3 of Chapter 4). In most well-studied diseases, a proportion of cases (though usually only a minority) are caused by partial or total deletion of the gene. Deletions which affect only a single gene are always far below the limit of cytogenetic detection but, if the gene has been cloned, deletions can be detected by the failure of a probe to hybridize to the DNA, or by the appearance of an abnormally sized fragment in a restriction digest. Pulsed field electrophoresis (see Section 8.1 of Chapter 3) is the best technique for finding abnormally sized fragments. Lack of hybridization is easy to see in males with X-chromosome deletions, but harder elsewhere because normally only one of the two homologous chromosomes carries the deletion. A decrease in band intensity may be noticeable on the autoradiogram if some bands are due to two copies of a sequence and others to only one. Dosage is easier to see if a second probe is used to give control bands of known gene dosage in each track, and the band intensities are measured using a scanning densitometer. Unfortunately, results are rarely reliable enough to be used in genetic counselling.

Direct tests can also detect point mutations as well as deletions. If the DNA sequence of the normal and mutant alleles is known, it is possible to use allele-specific oligonucleotides or ARMS on PCR-amplified DNA, as described in Section 3.3 of Chapter 1. If the normal sequence but not the mutant sequence is known (as often happens when somebody has an uncharacterized mutation in a well-characterized gene) it may be possible to sequence the DNA or use SSCPs or a heteroduplex assay (see Section 3.3.1 of Chapter 1) to find the unknown mutation. With large genes this is seldom practicable—for example, virtually no non-deletion dystrophin mutations have been described, although the cDNA has been available since 1987. Where direct tests are not possible, gene tracking can be used.

6.2 Gene tracking

Gene tracking uses linked markers to follow the transmission through a pedigree of a chromosomal section‹which is known to carry a disease gene. The prerequisite is that it is a Mendelian disease with a known map location, and one or more linked RFLPs. Gene tracking has the disadvantage that it requires family studies, and is applicable only if the pedigree structure is suitable; against this, it does not require any knowledge of the molecular pathology, and so can be used with diseases where the gene has not yet been isolated, or where the mutations are very heterogeneous. The investigation always follows the same three steps:

(a) Find a marker which is closely linked to the disease locus and for which the person at risk of transmitting the disease is heterozygous.

(b) Work out which of the marker alleles is on the chromosome carrying the pathological disease allele. This is done by studying other family members.

(c) Use the marker to discover whether the pathological chromosome or its normal homologue was passed on to the person requiring the diagnosis.

Figure 2.13 shows predictive testing in Huntington's disease (HD). Someone whose parent is affected has a one in two chance of carrying the gene, but symptoms do not usually appear until middle age. Some younger people want to know whether they carry the gene in order to make decisions about marriage, childbearing, and so on. At the time of writing the HD gene has not been cloned, but, as mentioned previously, it has been mapped to the short arm of chromosome

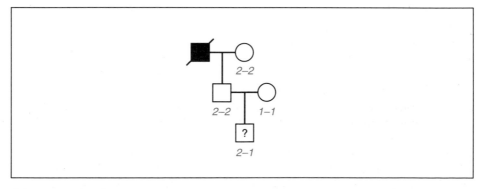

Figure 2.13. Gene tracking using a linked marker as a predictive test in Huntington's disease (HD). For definitions of the symbols, see *Figure 2.5*.

(d) The first step is to find a marker which maps to this location and for which the affected parent is informative. Phase is assigned by testing the surviving unaffected grandparent—she is 2 – 2, therefore the affected parent must have inherited allele 1 with the disease from his father (now deceased). In the third step, the patient and his mother are tested and this shows that the allele he has inherited from his affected father is allele 2. He has therefore inherited the grand-maternal allele, and is at low risk of developing HD. His risk is not zero, however, because there might have been a crossover during the paternal meiosis, which caused the disease gene to travel with marker allele 2.

This example shows the power but also the limitations of gene tracking. The pedigree structure must be suitable (it would not have been possible if both grand-parents were dead). The family must all agree to give blood for DNA analysis, even though the test does not benefit most of the people bled. Unacknowledged non-paternity can lead to completely wrong deductions, and it is a delicate question whether samples should be DNA fingerprinted (24) to check paternity, and if so whether the family should be told that this will be done. Most importantly, the marker types must come out in informative combinations. Successful gene tracking requires a number of markers, so that if one is uninformative another can be tried.

Recombination places a fundamental limit on the accuracy of gene tracking. In the Huntington's disease example, if the marker shows 4 per cent recombination with the disease, the prediction is only 96 per cent certain. The risk of a false negative result in prenatal diagnoses is especially worrying. There are rules of thumb that 1 per cent recombination corresponds to a distance of one million base pairs, and that one nucleotide in 200 is polymorphic. These generalizations suggest that it should be possible with any disease to find plenty of polymorphisms so close to the disease gene that recombination is negligible. In practice, some disease genes (for example, dystrophin) are at recombination hot spots, and with others the search for close markers has been very slow, so that in many cases the recombination problem is a real one.

One solution to the recombination problem is to use bridging (sometimes called flanking) markers. *Figure 2.14* shows an example of bridging markers in carrier detection for Duchenne muscular dystrophy. Two markers are used, one mapping either side of the disease locus. With the marker types as shown, the proband could be a carrier only if she is a double recombinant. If each marker shows 5 per cent recombination with the disease gene, then the risk of double recombination is 0.25 per cent and the proband can be 99.75 per cent sure that she is not a carrier. In practice, the probability of double recombination is even lower, because one crossover during meiosis tends to inhibit the formation of other crossovers close by, a phenomenon known as interference.

Had the proband in *Figure 2.14* turned out to be A*1 – 1 B*1 – 1, this would indicate that there had been a single recombination in the maternal meiosis which produced her, but without extra, closer markers, there is no way of knowing whether the recombination was proximal or distal to the disease gene. In these circumstances it would not be possible to make a prediction from the DNA studies. If the outside markers each show 5 per cent recombination with the

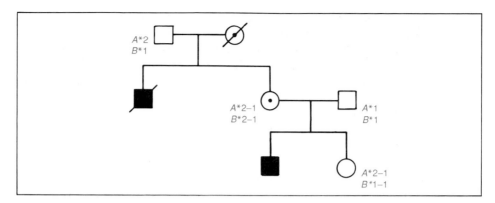

Figure 2.14. Gene tracking using bridging markers for carrier detection in DMD. (N.B. Males have only one allele of each marker because they have only one X-chromosome.) For definitions of the symbols, see *Figure 2.5*.

disease, recombinants between the markers are likely to occur in almost 10 per cent of cases. The gain in accuracy with bridging markers must be paid for by a reduction in the number of families where predictions can be made.

The ideal solution to the recombination problem is to use RFLPs within the disease gene itself. Normally these will be polymorphisms in introns. They are unrelated to the mutational event which converts the normal allele into a disease allele. You might think that if enough is known about the disease gene to make intragenic probes available, then it would be simpler to discover the actual disease-causing mutation and test for it directly. In many cases this will be true, but by no means all. For diseases which are heterogeneous at the molecular level, direct tests may be impracticable. For these cases, intragenic RFLPs offer the simplest and most reliable diagnosis.

7. Further reading

On human gene mapping:
Stephens,J.C., Cavanaugh,M.L., Gradie,M.I., Mador,M.L., and Kidd,K.K. (1990) *Science,* **250**, 237.

On the principles of linkage analysis:
Strickberger,M.W. (1976) *Genetics.* Macmillan, New York.
Snyder,L.A., Freifelder,D., and Hartl,D.l. (1985) *General genetics.* Jones and Bartlett, Boston.
Any other undergraduate general genetics textbook.

On linkage analysis in man:
Ott,J. (1991) *Analysis of human genetic linkage (2nd edn),* Johns Hopkins University Press, Baltimore. (This is an essential practical guide for anyone wishing to do it.)
White,R., Leppert,M., O'Connell,P., Nakamura,Y., Julier,C., Woodward,S., Silva,A., Wolff,R., Lathrop,M., and Lalouel,J.M. (1986) *Construction of human genetic linkage maps:*

1. *Progress and perspectives. Cold Spring Harbor Symposia on Quantitative Biology, L1, 29.* Cold Spring Harbor Laboratory, New York.

Lalouel,J.-M., Lathrop,G.M., and White,R. (1986) *Construction of human genetic linkage maps: II. Methodological issues. Cold Spring Harbor Symposia on Quantitative Biology LI, 39.* Cold Spring Harbor Laboratory, New York.

Lander,E.S., and Botstein,D. (1986) *Mapping complex traits in humans: new methods using a complete RFLP linkage map. Cold Spring Harbor Symposia on Quantitative Biology LI, 49.* Cold Spring Harbor Laboratory, New York.

On human genetic disease:

Connor,J.M. and Ferguson-Smith,M.A. (1991) *Essential medical genetics* (3rd edn), Blackwell, Oxford.

Gelehrter,T.D. and Collins,F.S. (1990) *Principles of medical genetics.* Williams & Wilkins, Baltimore.

8. References

1. ISCN (1985) *An international system for human cytogenetic nomenclature.* Karger, Basel.
2. Harper,M.E. and Saunders,G.F. (1981) *Chromosoma*, **83**, 431.
3. Lichter,P., Tang,C.-J.C., Call,K., Hermanson,G., Evans,G.A., Housman,D., and Ward,D.C. (1990) *Science*, **247**, 64.
4. Trask,B., Pinkel,D., and van den Engh,G. (1989) *Genomics*, **5**, 710.
5. Ruddle,F.H. (1981) *Nature*, **294**, 115.
6. Theune,S., Fung,J., Todd,S., Sakaguchi,A.Y., and Naylor,S.L. (1991) *Genomics*, **9**, 511.
7. Cavanee,W.K., Dryja,T.P., Phillips,R.A., Benedict,W.F., Godbout,R., Gallie,B.L., Murphree,A.L., Strong,L.C., and White,R.L. (1983) *Nature*, **305**, 779.
8. Vogelstein,B., Fearon,E.R., Kern,S.E., Hamilton,S.R., Preisinger,A.C., Nakamura,Y., and White,R. (1989) *Science*, **244**, 207.
9. Foy,C., Newton,V.E., Wellesley,D., Harris,R., and Read,A.P. (1990) *Amer. J. Hum. Genet.*, **46**, 1017.
10. Grant,S.G. and Chapman,V.M. (1988) *Ann. Rev. Genet.*, **22**, 199.
11. Brown,C.J., Ballabio,A., Rupert,J.L., Lafreniere,R.G., Grompe,M., Tonlorenzi,R., and Willard,H.F. (1991) *Nature*, **349**, 38.
12. Emery,A.E.H. (1986) *Methodology in medical genetics*, (2nd edn), (Chapter 4). Churchill Livingstone, Edinburgh.
13. Byers,P.H. (1989) *Amer. J. Med. Genet.*, **34**, 72.
14. McKusick,V.A. (1991) *Mendelian inheritance in man*, (10th edn). Johns Hopkins University Press, Baltimore.
15. Morton,N.E. (1955) *Amer. J. Hum. Genet.*, **7**, 277.
16. Ott,J. (1974) *Amer. J. Hum. Genet.*, **26**, 588.
17. Lathrop,G.M., Lalouel,J.M., Julier,C., and Ott,J. (1984) *Proc. Natl. Acad. Sci. USA*, **81**, 3443.
18. Connor,J.M. and Ferguson Smith,M.A. (1991) *Essential medical genetics,* (3rd edn) (Appendix II, p. 240). Blackwell, Oxford.
19. Read,A.P., Thakker,R.V., Davies,K.E., *et al.* (1986) *Human Genetics*, **73**, 267.
20. Edwards,J.H. (1987) *J. Med. Genet.*, **24**, 539.
21. Weeks,D.E. and Lange,K. *Amer. J. Hum. Genet.*, **42**, 315.
22. Ardinger,H.H., Buetow,K.H., Bell,G.I., Bardach,J., Vandemark,D.R., and Murray,J.C. (1989) *Amer. J. Hum. Genet.*, **45**, 348.
23. Ivinson,A., Read,A.P., Harris,R., *et al.* (1989) *J. Med. Genet.*, **26**, 426.
24. Jeffreys,A.J., Wilson,V. and Thein,S.L. (1985) *Nature*, **316**, 76.

3

The identification of genes in human inherited disease

1. Introduction

Since the first mammalian gene was cloned in the early seventies, many genes have been isolated that are known to be involved in the pathology of human diseases. The approaches used for their isolation have depended on the resources available and the levels of expression of the genes concerned. In this chapter, the main methods are described both for diseases where the gene involved is known in advance and for those disorders where the strategy has necessitated a more indirect approach to be adopted. In the latter situation, the chromosomal location of the disease gene is known. The gene is isolated by reverse genetics approaches, often referred to as 'positional cloning'.

2. mRNA enrichment of specific mRNA populations

2.1 Populations of mRNA in eukaryotic cells

The mRNA population in a typical eukaryotic cell is made up of between 10 000 and 30 000 different species present at several abundance levels. These have been estimated by hybridization studies to fall into three classes (1) approximately as follows:

    30   different species at 3500 copies per cell
    1090   different species at   230 copies per cell
    10670 different species at   14 copies per cell

Obviously, the strategies used for the isolation of a particular gene sequence will be influenced greatly by its relative abundance in the cell. In the case of most inherited disorders, the gene responsible is present at low abundance which explains why differences related to the phenotype are not detected by two-dimensional protein gel electrophoretic analyses. The advent of PCR has made cDNA cloning very much easier; cDNA libraries of high complexity can be constructed from very small quantities of mRNA.

2.2 Abundant sequences

The globin genes were cloned by virtue of the fact that their transcripts are the main mRNA species in erythroid cells. Thus if a cDNA library (2) constructed from polyA$^+$ mRNA from this tissue is screened with mRNA from erythroid cells most of the positive clones would correspond to copies of globin mRNA. Abundant tissue-specific sequences can be easily identified by screening a cDNA library made from the tissue where the sequence is of high abundance with polyA$^+$ mRNA from the same tissue and with polyA$^+$ mRNA from a tissue where the sequence is either absent or present in low abundance. This strategy can be applied to mRNA present at levels as low as 0.05 per cent of the total mRNA in the cell.

2.3 Low abundance sequences

The isolation of low abundance genes often depends on the enrichment of the polyA$^+$ mRNA by subtractive hybridization prior to differential screening of cDNA libraries.

2.3.1 T cell receptor gene

Hedrick and colleagues exploited subtractive hybridization very elegantly to isolate the T cell receptor gene (3). The strategy depended upon four assumptions about the nature of the gene:

(a) that it should be expressed in T cells but not in B cells;

(b) that the mRNAs for the T cell receptor proteins should be found on membrane-bound polysomes, as one would expect the nascent receptor polypeptides to attach to the endoplasmic reticulum by a leading peptide, or signal sequence;

(c) that, like immunoglobulin genes, those genes that encode the T cell receptor proteins should be rearranged in T cells;

(d) that, like immunoglobulin genes, they should have constant and variable regions.

As B and T lymphocytes differ in only a small fraction of their gene expression (200 – 300 different sequences) and only a small proportion of mRNAs of lymphocytes appear to be in the membrane-bound polysomal fraction, the experimental strategy given in *Figure 3.1* was adopted. By synthesizing ^{32}P-labelled cDNA from membrane-bound polysomal RNA of antigen-specific T cells and removing those sequences also expressed in B cells by subtractive hybridization, the specific 'enriched' probe could then be used to screen a library of cloned cDNAs made from another selected library. Positive clones were identified as encoding the T cell antigen receptor since they showed somatic gene rearrangement and revealed variable and constant regions.

The application of subtractive hybridization techniques to the analysis of genes

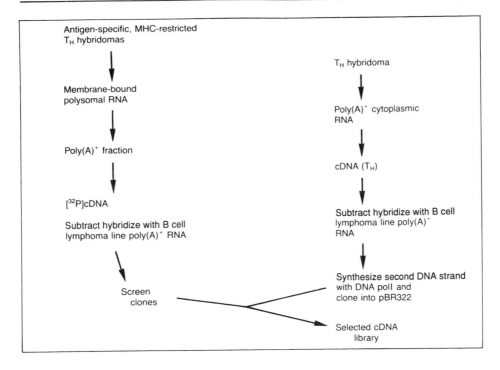

Figure 3.1. Strategy for cloning T cell antigen receptor.

is now much less complex. This is because a technique has been developed to attach specific DNA sequences to magnetic beads, hybridize them in a solution containing the desired sequence or sequences, and then separate out the beads with their relatively pure hybridized sequences in a magnetic field.

2.3.2 Chronic granulomatous disease gene

Chronic granulomatous disease (CGD) was also identified after subtractive hybridization of mRNA populations. This was also the first example of a human disease mutation being isolated from its chromosomal location ('positional cloning'; 4). CGD is X-linked and affected boys have impaired function of the oxidase system in their phagocytic cells (neutrophils, macrophages, and eosinophils) and suffer from recurrent, severe bacterial and fungal infections. Although biochemical studies by Segal and colleagues (5) had suggested that the basic defect might lie in cytochrome *b* since all X-linked CGD patients lacked this protein, there were conflicting data in the literature.

The mutation in the X-linked CGD gene was first localized to Xp21 by the cytogenetic detection of affected boys and carrier females with deletions in this region (6,7); this was later confirmed by linkage analysis (8). Royer-Pokora *et al.*, (4) had already isolated an enriched set of sequences for the candidate gene by anticipating its elevated expression in normal mature neutrophils. An enriched cDNA population was isolated by subtractive hybridization of cDNA, made from mRNA prepared from induced human leukaemic HL60 cells, with mRNA, prepared from cultured B lymphocytes derived from a patient with a deletion in Xp21 who suffered from CGD and Duchenne muscular dystrophy. These sequences were then used to screen an enriched genomic library derived from a deletion in a second CGD/DMD patient with a larger deletion who also suffered from retinitis pigmentosa and showed the McLeod phenotype (see Section 7.2 and *Figure 3.2*). Two genomic clones identified by this procedure were used to screen a cDNA library from an induced HL60 cell line. A transcript of 5 kb was identified in northern blots and subsequently cloned.

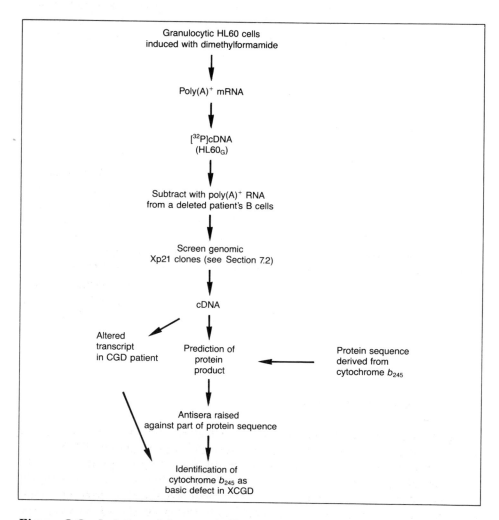

Figure 3.2. Isolation of the gene in X-linked CGD.

Evidence that this transcript was indeed the one mutated in CGD patients was derived from the observation that non-deleted patients showed none or reduced levels of the transcript and one patient produced a truncated mRNA molecule. The latter result is the critical one in relating the clone to the disease.

The function of the molecule was elucidated partly from its amino acid sequence, deduced from the DNA sequence. The molecule showed no homology to previously described cytochromes and contained a large hydrophobic region that might function as a transmembrane domain. Further analysis of the sequence suggested that the protein might be a membrane glycoprotein. Antibodies were raised against a synthetic peptide derived from the cDNA sequence and a protein with a molecular weight of 90 000 was identified in neutrophils. These antisera were shown to react with the 90 kd component of the purified subunit of the cytochrome b_{245} (9). Correspondence was also found between the small amount of amino acid sequence and the protein sequence deduced from the cDNA (10). Thus, both conventional biochemistry and reverse genetics were important in the discovery of the basic defect in X-linked CGD.

3. Complementation and gene rescue

If an appropriate tissue culture line is available which is mutated at the locus of interest, then the gene can be identified by introducing DNA into the cell line and assaying for clones where the defect is corrected. This is essentially the approach which was profitably applied for the cloning of the hypoxanthine guanine phosphoribosyl transferase (HPRT) gene. Human genomic DNA was transfected into an HPRT-deficient mouse cell line and the resulting transfectants grown on selective medium to determine whether any of them had incorporated human genomic DNA sufficient to complement the mutation (11). The DNA sequences in the selected transfectants could be rescued by making a library and screening for human AluI repeat sequences, which would have been co-integrated into the cell line together with the human genomic DNA encoding the HPRT gene.

4. Expression screening

In many instances, polyclonal or monoclonal antibodies are available for the protein coded by the probe sequence of interest. In these cases a cDNA library cloned into an expression vector can be screened for the gene product directly. It is usually produced in bacterial clones as a fusion protein with the β-galactosidase gene (2). Libraries of very high complexity (several million recombinants) can be readily cloned enabling the identification of even very low abundance class genes.

5. Oligonucleotide screening

In some instances, although an antibody is not available, it may be possible to obtain enough information to deduce a small amount of the sequence of the protein. This information can then be used to synthesize a series of corresponding oligonucleotide probes. A length of at least 11 nucleotides is required for specific hybridization and the dissociation temperature (T_d) of the hybrids (the temperature at which half of the duplexes are dissociated) can be calculated from the formula:

$$T_d \ (°C) = 4(G + C) + 2(A + T)$$

where G,C,A and T indicate the number of the corresponding nucleotides of the oligomer. For example, the T_d of the sequence TGACTCCGTAA would be 32°C.

This is a very widely applicable strategy for the cloning of human genes (12) and has been successfully used in many instances, notably for the cloning of Factor VIII, $\alpha-1-$antitrypsin and the LDL receptor gene.

6. Cloning of chromosome breakpoints

For certain genetic disorders, cytogenetic analysis of a patient's chromosomes has led to a strategy for the cloning of candidate genes. The classic example of this is the cloning of sequences from within the Duchenne muscular dystrophy gene by the cloning of a translocation breakpoint (13).

Duchenne muscular dystrophy (DMD) is an X-linked muscle wasting disorder affecting approximately one in 3000 males. Affected boys usually have to use a wheelchair before the age of 12 years and die in their early twenties. Females are not normally affected but a few have been reported where the phenotype is quite severe. Cytogenetic analysis of some of these females reveals an X; autosome balanced translocation where the breakpoint on the X-chromosome is within Xp21. Different autosomes are involved in each case. In X;autosome balanced translocations it is the translocated X-chromosome that is active and the normal X-chromosome which is preferentially inactivated (see Section 2.4 of Chapter 1). Thus, if Xp21 is the site for the DMD gene, then this would explain the manifestation of the disorder in these individuals.

One particular translocation offered the possibility of cloning the translocation breakpoint itself and therefore a unique strategy for isolating the gene involved in DMD (13). This was the translocation between the X-chromosome and chromosome 21, where the breakpoint on the latter lay within a cluster of rDNA genes (see schematic presentation in *Figure 3.3*). If a library of the derivative X-chromosome could be made, screening of the clones with a rDNA probe should lead to the identification of a recombinant which contains not only rDNA sequences, but also X-chromosome sequences.

Worton and colleagues first characterized the X;21 breakpoint by restriction enzyme analysis and Southern blotting since initial attempts to clone a library

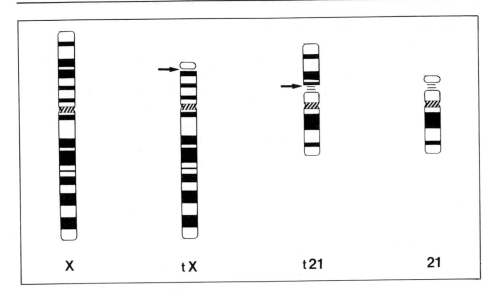

Figure 3.3. Balanced translocation in a female suffering from muscular dystrophy.

with the junction fragment proved unsuccessful. By using a rDNA sequence probe, they identified a *Bam*HI fragment present on the derivative X-chromosome in a hybrid cell line made from the patient and not normally present in human DNA. They cloned this fragment into a bacteriophage vector after excising it from an agarose gel (for examples of this type of procedure see ref. 2). Hybridization of this clone later confirmed that this sequence contained material homologous not only to rDNA but also to X-chromosome sequences within the Xp21 region (13). It was given the laboratory acronym XJ and the derivative subclone used to screen patients for deletions XJ1.1. XJ1.1 was found to be deleted in 10 per cent of DMD patients and has subsequently been shown to lie within one of the large introns of the DMD gene (see Section 3.3 of Chapter 4).

Sequencing across the chromosome breakpoint in this patient has revealed that a small deletion of DNA occurred during the translocation event. No obvious consensus sequence near the junction was found on either chromosome to indicate the molecular mechanism underlying the exchange.

7. Genomic enrichment

Once a genetic disorder has been localized to a chromosomal region either by linkage studies with RFLPs or deletion mapping, then the first step in the identification of the genes associated with the disease is the selection of sequences from the region of that particular chromosomal sub-band. There are at least five main strategies which can be adopted in this process.

7.1 *Chromosomal sorting*

Most genetic disorders are not associated with a deletion or translocation break-point. The defective gene is localized to a chromosomal region by linkage analysis and flanking markers usually define the disease locus to within approximately 10–20 megabases of DNA. One of the strategies to identify closer markers is to localize clones picked at random from a chromosome enriched library. The clones are mapped to the region of interest using somatic cell hybrids. The chromosome enriched libraries are constructed from small amounts of DNA purified by chromosome sorting.

The principle of chromosome sorting is shown in *Figure 3.4*. Essentially, metaphase chromosomes, stained with an appropriate fluorescent dye such as

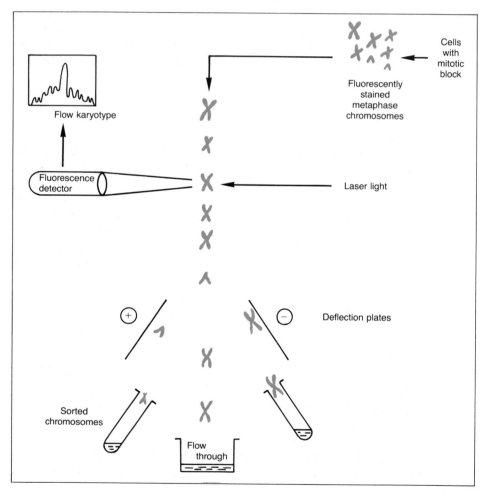

Figure 3.4. Principle of chromosome sorting.

ethidium bromide, are passed through a laser beam. The resulting fluorescence emission of the various chromosomes will be dependent upon the amount of dye bound, which in turn is proportional to the DNA content (and hence size) of the chromosome in the case of ethidium bromide. A series of detectors can monitor and sort the chromosomes which are associated with the correct amount of fluorescence. Better resolution can be obtained if two dyes (such as chromomycin and Hoechst) are used and two laser beams utilized. This technique has become so powerful that it is now used to localize sequences within the human genome. Chromosomes are collected directly on to filters for hybridization or into separate pots for the localization of sequences of interest by PCR amplification. The position of a chromosome in the fluorescence profile is very sensitive to DNA content thus making this technique also useful for sorting aberrant chromosomes and for karyotyping.

Chromosome sorted libraries are an important resource for mapping disease mutations and are generally available. There are now both bacteriophage and cosmid libraries for all the human chromosomes. Some of the cosmid libraries have been gridded out as individual clones in microtitre dishes for the construction of total physical maps of human chromosomes (see Section 6 of Chapter 4).

7.2 Reassociation techniques

Genomic reassociation techniques were initially developed to isolate sequences from the human Y-chromosome. They have been extended to the analysis of the Duchenne muscular dystrophy locus by Kunkel and colleagues, who demonstrated the power of this approach for the analysis of X-linked disorders (14). The principle of the strategy is depicted in *Figure 3.5*.

DNA from a patient with a deletion is sonicated to a size of about 500 bp and heat denatured. DNA from a normal, non-deleted individual, or in this case a 48,XXXXY individual to enrich for X-chromosome sequences, is cut with a restriction enzyme such as *Sau*3A to give defined fragments of average size 500 bp and heat denatured. These two populations of single-stranded DNA fragments are then mixed in the ratio of 200:1 (deleted:non-deleted) and allowed to reassociate. Three types of hybrids will result:

(a) hybrids where sonicated, deleted DNA has reassociated to itself;

(b) hybrids where sonicated, deleted, and non-deleted sequences have reassociated;

(c) hybrids where the *Sau*3A fragments from the 48,XXXXY DNA have reassociated together.

Of these groups, only the third type of hybrids are clonable into a *Bam*HI vector, since only these hybrid molecules will have two defined, precisely cut ends. In addition, since the deleted sonicated DNA is in large excess, the third type of hybrids will be predominantly derived from sequences that cannot be competed out by the sequences from the deleted individual, that is, those from the deleted region.

Kunkel and colleagues exploited this technique using a patient suffering from

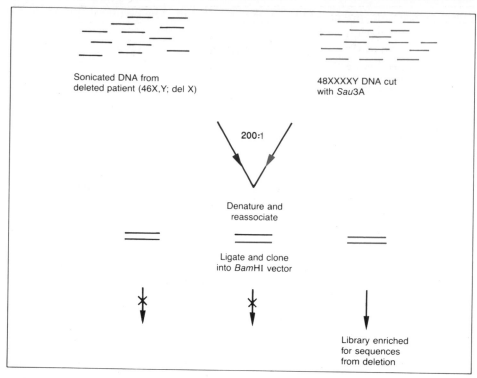

Figure 3.5. Genomic reassociation.

DMD, CGD, retinitis pigmentosa, and the McLeod phenotype, who had a cytogenetically detectable deletion within Xp21. The hybridization reaction was carried out in a phenol emulsion to accelerate the rate of reassociation so that the vast majority of single-copy sequences had reassociated. The sequences have become known as phenol-enhanced reassociation technique (PERT) sequences. One of these, PERT87, was shown to be deleted in not only the original patient used in the experiment but also in 10 per cent of patients that suffer only from DMD. Expansion of this sequence by chromosome walking finally led to the identification of the DMD gene (see Section 3.3 of Chapter 4).

7.3 Chromosome-mediated gene transfer

Chromosome-mediated gene transfer (CMGT) is a very powerful approach for the preparation of 'minigenomes' from the human karyotype in a rodent background (see ref.15 for review). It has been used notably in the search for genes associated with Wilm's tumour on chromosome 11p and cystic fibrosis on chromosome 7q.

The experimental strategy used to make the CMGT hybrid lines is shown in *Figure 3.6*. The isolation of transformants depends on the availability of a

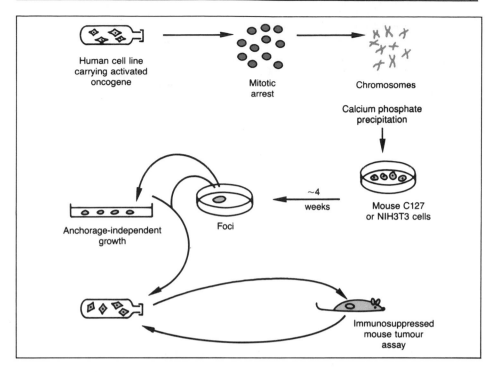

Figure 3.6. Construction of CMGT hybrid lines.

selectable marker near the region of interest. In the case of Wilm's tumour, H-*ras* was used and the *met* oncogene was exploited for cystic fibrosis. If such selection is not available, then the human chromosome fragments may be co-transfected with a selectable marker (such as the *neo* gene which encodes resistance to the antibiotic Geneticin G418 in eukaryotic cells) when it is co-integrated into the DNA.

More recently, useful hybrid lines have been produced after fragmentation of the chromosomes by irradiation (so-called irradiation hybrids; 16). A somatic cell hybrid containing the human chromosome of interest as its only human component is irradiated and the recipient rodent line is chosen so that selection can be applied for fusion hybrids between the new recipient cell and the irradiated donor. The uptake of human chromosome fragments into the rodent chromosomes is random but appropriate lines can be obtained by assaying them for the presence of DNA sequences lying in the region of interest.

Alternatively, monoclonal antibodies can be used to select for the expression of cell surface markers in the transfected cells. Those expressing the marker can be isolated by sequential rounds of enrichment using a fluorescence activated cell sorter (FACS; ref.17).

Irradiation hybrids are being constructed for the analysis of many human genetic disorders. The one disadvantage is that, for reasons as yet unknown, fragments containing parts of the centromere are nearly always present in the transformants and there may be more than one human fragment integrated into two different rodent chromosomal regions. The co-linearity of the integrated DNA and the absence of substantial rearrangements of DNA sequences can be monitored by blotting the transfectants against a medium abundance repeat such as the L1 repeats (18). *Figure 3.7* shows an example of this analysis, carried out

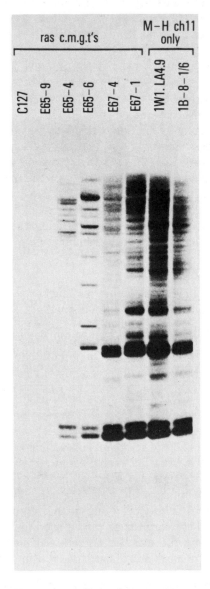

Figure 3.7. L1 fingerprinting. The different CMGT lines show different patterns (taken from ref. 18 with kind permission).

on chromosomes HRAS1-selected CMGTs along with two mouse – human hybrid lines. Thus, the L1 fingerprints of E67 – 1 and E67 – 4 are similar, both carrying subsets of the complete chromosome 11 pattern. On the other hand, E65 – 6 shows evidence of local rearrangement since the pattern seen is quite dissimilar to that of the other transformants.

The overlap in human genomic content of various CMGT lines can also be assayed by *Alu*-PCR (19). As the name suggests, this technique takes advantage of the fact that *Alu* sequences occur often enough in the human genome such that primers made from their consensus sequence can be used to amplify the sequences between the *Alu* repeats. The pattern of *Alu*-PCR products can be used as a fingerprint for the human DNA contained within the hybrid. This *Alu*-PCR profile can be very useful if hybrids need to be checked regularly and can avoid the necessity to grow the lines up for time-consuming karyotype analysis, *Alu*-PCR can also provide the starting point for the isolation of sequences from the region of human DNA contained in the hybrid (see Section 7.4).

The quality of a particular irradiation hybrid line can be assessed by a technique known as chromosome painting (20). This involves preparing DNA from a chromosome-sorted library, labelling it with a fluorescent marker and hybridizing the DNA (in the presence of competitor to suppress hybridization of repetitive DNA) to a metaphase spread of the hybrid. The hybridization obtained is compared with the result obtained using *Alu* sequence as labelled marker which will detect all human DNA. A good hybrid will show a single site of fluorescence corresponding to the one integrated human fragment.

7.4 Alu-PCR

Alu-PCR is a routine method of obtaining sequences from a chromosomal region of interest. If a suitable hybrid containing a small enough human fragment has been constructed by irradiation, *Alu*-PCR can be used to identify the specific human sequences (19). The PCR products can be either cloned directly or radio-actively labelled and used to screen genomic libraries. *Alu*-PCR is used extensively to generate probes for the screening of yeast artificial chromosomes (YAC) and cosmid libraries in order to construct complete physical maps of chromosomal regions (see Section 8.4; 21).

7.5 Microdissection

Microdissection of chromosomes was initially developed for the analysis of the mouse T complex. However, it did not become a routine procedure until the advent of PCR. The combination of microdissection with PCR enables very small amounts of material to be cloned and hence the construction of libraries specific for small regions of the genome (22). Several regions important in human diseases have been targeted in this way, including the Prader – Willi region on chromosome 15q, the Wilm's tumour region on chromosome 11p, and the fragile X-region at Xq27.

The chromosomal segment is excised with a needle using a micromanipulator. Alternatively, the chromosome region of interest can be selected by melting away

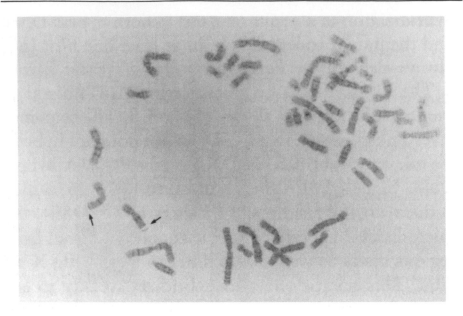

Figure 3.8. Microdissection of Xq27 (with kind permission of *Am. J. Hum. Genet.*).

all other chromosomal regions using a laser. The dissected material is digested with the restriction nuclease *Rsa*I so that it can be cloned into the blunt ends of a suitable vector. *Rsa*I is chosen because it does not cut at *Alu*I sequences and therefore should bias the library away from highly repetitive sequences. The DNA is then amplified using the vector sequences as primers. The PCR products are then cut with *Eco*RI and cloned into the *Eco*RI site of a second vector. The region that can be targeted is usually of the order of 10–20 Mb of DNA. An example of the needle microdissection of the Xq27 region is shown in *Figure 3.8*.

The combination of YAC cloning (see Section 8.5) and microdissection is a very powerful approach to the construction of a physical map of the human genome. Microdissection of Xq27 has enabled the isolation of the total genomic DNA of the region (23). Each microclone of about 200 bp in length can be sequenced to provide PCR primers (sequence tagged sites, STSs) for the screening of YAC libraries for YAC clones covering 500 kb or more.

8. Identification of candidate genes

8.1 Pulsed field electrophoresis

The lower level of resolution for chromosome banding is of the order of 3000 kb whereas the upper level of resolution for conventional electrophoresis is 20–30 kb. Pulsed field gel electrophoresis (PFGE) bridges the gap between these two methods of physical mapping because it is capable of separating fragments from 50 kb up to 10 000 kb (24; for review, see ref. 25). This technique is widely used to construct a detailed map of the region containing the disease locus of interest. In many cases, PFGE has been used to identify deletions and/or translocations associated with a phenotype, thereby pinpointing the position of the defective gene.

Large DNA fragments are separated in PFGE because the DNA molecules are subjected, alternately, to two approximately perpendicular fields. The separation is based on the fact that when molecules in a gel matrix are subjected to an electric field perpendicular to their direction of migration, they will reorientate themselves along the new field. Small molecules will be able to do this more readily than large ones. The resolution depends on the time of switching between the two fields (pulse time). An example of the separation is shown in *Figure 3.9*. The gel shows the separation of human DNA digested with two infrequently cutting restriction enzymes together with bacteriophage lambda concatemers (multiples of 50 kb) as markers. The desired resolution can be obtained by changing the pulse times: longer pulse times will resolve DNA fragments in the higher molecular weight range. The DNA samples separated by PFGE can be Southern blotted on to a membrane and hybridized in exactly the same way as conventional gels. The large DNA molecules can be broken down for transfer by acid treatment or UV irradiation of the DNA in the gel.

Modifications of the original PFGE method of Cantor and Schwartz (26) described above have been developed which result in the DNA running straight in the tracks thus improving the resolution. The field is generated by the use of contour-clamped homogeneous electric fields (CHEF) and the apparatus consists of an hexagonal array of electrodes to create a more homogeneous field (27). The resolving power can then be optimised for fragments from 1 – 10 Mb.

The cytosine residues in the dinucleotide CpG are the major site of methylation in eukaryotes and the methylated C residue is frequently deaminated to thymine leading to an under-representation of CpG in the genome (27). Thus enzymes used to cut human DNA into large DNA fragments are those which have the dinucleotide CpG in their recognition site. Furthermore, these restriction enzymes only cleave at their target sites if they are unmethylated. Therefore the general distribution of cleavable sites within a region provides an estimate of both G + C content and methylation status of the DNA. Some of the restriction enzymes commonly used in PFGE are listed in *Table 3.1*.

A long-range restriction map of a chromosomal region can be generated by the same principle as conventional restriction mapping (28). Physical linkage between two markers can be established if at least two restriction fragments are identical in size and a detailed map can be built up from information on single and double digests. Partial digests can also be used to generate maps over much larger regions and to link more widely-spaced DNA probes. This approach is often used in PFGE analyses because the enzymes used are methylation-sensitive and partial digests are seen even after extensive digestion. PFGE can be used to identify candidate genes and assist in their isolation if the disease is caused by DNA deletion in some patients. In these cases, a change in the restriction

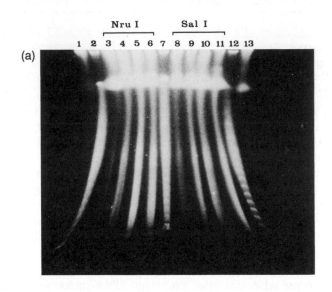

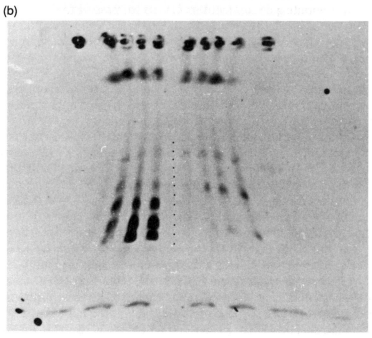

Figure 3.9. Separation of human DNA cut with enzymes *Nru*I and *Sal*I by PFGE. (a) Ethidium bromide stained gel; (b) hybridized to probe 2bC6 after blotting. Tracks 2, 7, and 12 are bacteriophage concatemers as markers.

Table 3.1 Commonly used, infrequently cutting, restriction enzymes

Enzyme	Recognition sequence
*Bss*HII	GCGCGC
*Eag*I	CGGCCG
*Mlu*I	ACGCGT
*Not*I	GCGGCCGC
*Nru*I	TCGCGA
*Sac*II	CCGCGG
*Sal*I	GTCGAC
*Sfi*I	GGCCNNNNNGGCC
*Sma*I	CCCGGG
*Xho*I	CTCGAG
*Xma*III	CGGCCG

map can be detected when the DNA probe is 1 Mb or more away provided that there is an appropriate gel fragment that covers the probe and the deleted region. Once the deletion has been identified, a library enriched for sequences from within the region can be constructed from the normal fragment by gel elution and cloning. The clones can then be used to screen northern blots or cDNA libraries in order to identify candidate genes. If the deletion maps only a few hundred kilobases away from the DNA probe then chromosome jumping strategies or linking libraries can be used to move closer to the gene of interest (see Section 8.3).

The construction of long-range physical maps of the human genome should allow a direct comparison of physical and genetic distance. This comparison is already revealing 'cold' and 'hot' regions of recombination which should eventually lead to the identification of the sequences which determine the regulation of recombination in humans.

The detection of deletions by PFGE is proving to be particularly useful in the diagnosis of carriers in Duchenne muscular dystrophy (29). Because the locus is so large and gene dosage is not always reliable, the detection of a changed band is an unambiguous method of determining the carrier status of individuals. Around 70 per cent of DMD patients show deletions within the 2.3 Mb of the gene sequence.

PFGE can also be used to map chromosomal rearrangements such as translocations. Translocation breakpoints involving oncogenes in certain human malignancies have been shown to be heterogeneous. Long-range physical mapping of the breakpoint region by PFGE is an efficient strategy for the understanding of how these chromosomal rearrangements lead to the perturbation of the expression of oncogenes which results in the malignancy.

8.2 HTF islands

HTF islands are clusters of unmethylated *Hpa*II sites (30). They represent about 1 per cent of vertebrate genomes and were detected as a fraction of tiny fragments produced by cleavage of total genomic DNA with the methylation-sensitive enzyme *Hpa*II. The recognition site of *Hpa*II is CCGG, and these islands therefore lie in regions of relatively high G + C content. In fact, they are not CpG-rich; rather, the bulk of non-island DNA is CpG deficient due to the fact that these other residues are methylated and often deaminated to thymine. The only circumstances in which the islands have been found to be methylated is on the inactive X-chromosome. The islands are usually 1 – 2 kb long and there are about 30 000 of them dispersed throughout the haploid genome of mammals.

Extensive studies of HTF islands have demonstrated that they are associated with some genes, particularly with the 5′ ends of genes. All housekeeping genes so far sequenced that are transcribed by RNA polymerase II have HTF islands at their 5′ ends. Many tissue specific genes such as *Thy*-1 in the mouse and α-globin in man are also associated with HTF islands. The exact function of these islands remains to be determined, but they are likely to represent protein binding sites associated with gene transcription. The reader is referred to Bird's review for a more detailed discussion (30).

The high G + C content of HTF islands makes them sites for the rare cutting enzymes used in PFGE analyses (see *Table 3.1*). Thus, clustering of these sites can be indicative of an HTF island and gene sequence. Studies of cosmids with *Sac*II sites, for example, have demonstrated that three out of four *Sac*II sites are associated with HTF islands and genes as assayed by hybridization to northern blots of HeLa cell mRNA (31).

A useful cloning strategy which allows the identification of sequences containing potential HTF islands has been employed in the search for the cystic fibrosis gene. Estivill *et al.* (32) constructed a cosmid library of a reduced hybrid by cloning into the *Xma*I/*Hin*dIII site of a particular cosmid vector. Since the recognition site of *Xma*III (CGGCCG) overlaps with that of *Not*I (GCGGCCGC), many of the clones will be derived from CpG-rich regions. This strategy led to the isolation of a highly conserved gene sequence associated with an HTF island that lies very close to the cystic fibrosis locus.

8.3 Jumping and linking libraries

If the region in which the disease mutation has been identified by linkage or the presence of a deletion or translocation in patients, there are two important chromosome jumping methods which can be utilized to move from the DNA marker to the defective sequence several hundred base pairs away. These require the construction of either a jumping or linking library.

A chromosome jumping library involves the cloning of sequences separated by 100 – 250 kb of DNA (33). Genomic DNA is cut into fragments of this size by partial digestion, and isolated after separation by PFGE. The resulting linear molecules are then ligated into circles (*Figure 3.10*). This step is carried out at very low DNA concentrations to promote intramolecular joining over inter-

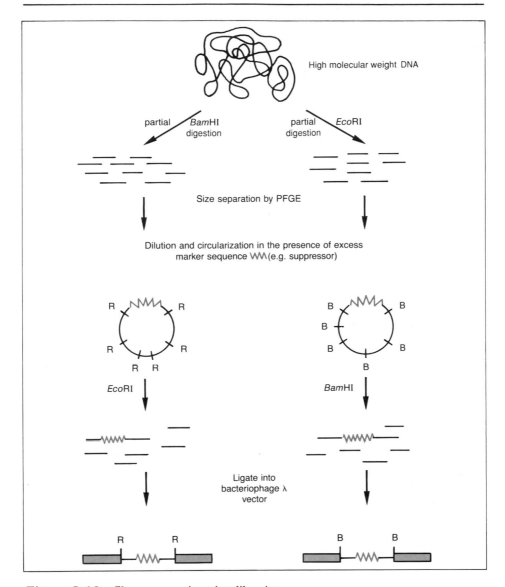

Figure 3.10. Chromosome jumping libraries.

molecular joining. This ligation reaction is normally carried out in the presence of an excess of a selectable genetic marker such as a suppressor gene (*sup*F). The circular molecules are then cut with a restriction enzyme other than that used to produce the original partial digestion of the genomic DNA. The end clones of the circles can then be cloned away from the sequences within the circles. To do this, the fragments are introduced into an appropriate bacteriophage vector carrying amber mutations and plated on a bacterial host not carrying a suppressor. This ensures that only recombinants containing the *supF* gene can grow. Screening this library with 'starter' sequences should then permit the identification of a sequence 100–250 kb DNA away. Furthermore, because the relative orientation of the restriction enzyme sites can be determined, the direction of the jumps can be deduced (see *Figure 3.11*). By screening different libraries made from different combinations of restriction enzymes, several directional jumps can be made (34, 35). The extent of these chromosomal jumps is limited by the fact that molecules larger than 250 kb do not circularize efficiently. Also, libraries of at least three million clones need to be constructed to have a 95 per cent probability of finding a clone for the next step. Even larger libraries may be needed if a significant number of the endpoints contain repetitive sequences.

Jumping libraries have also been described which have been constructed using rare-cutting enzymes such as *Not*I rather than partial digests. However, it is now clear that *Not*I sites are not randomly distributed throughout the human genome and some large fragments would not circularize efficiently. This problem can be overcome to a certain extent by using combinations of enzymes such as *Not*I complete digestion and partial *Eco*RI digestion, followed by circularization in the presence of a plasmid cleaved with both enzymes.

A more useful way of cloning fragments surrounding rare-cutter restriction enzyme sites that also allows directional jumping is the screening of a linking library in conjunction with a rare-cutter jumping library (35). The strategy is summarized in *Figures 3.12* and *3.13*. For the linking library, genomic DNA is partially cleaved with *Sau*3A and fragments of 10–20 kb collected after rate-zonal centrifugation on sucrose gradients. This DNA is then circularized at low concentration in the presence of a *Bam*HI-cut plasmid containing a suppressor gene. The circles are then cleaved with *Not*I and cloned into an appropriate bacteriophage vector (e.g. *Not*I cleaved EMBL3). Link clones are obtained by plating on a suppressor-free host. If small chromosome regions can be saturated with link clones, a physical map based on *Not*I fragments can be constructed

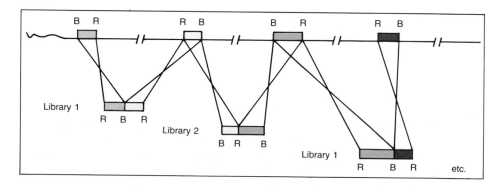

Figure 3.11. Directional jumping with two jumping libraries.

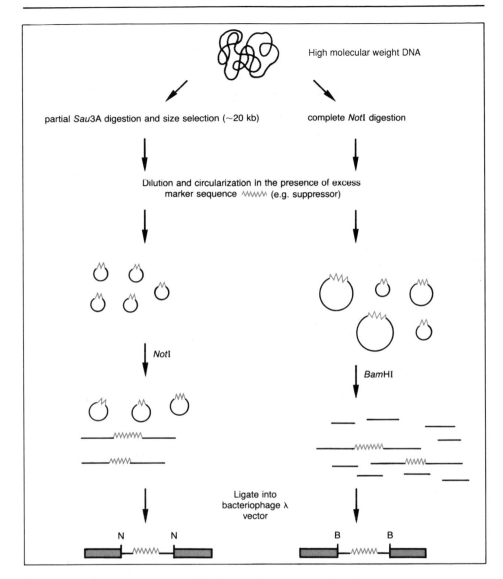

Figure 3.12. Construction of linking and rare-cutter libraries.

using PFGE. This strategy has been used for the mapping of the region surrounding the Huntington's gene and the T complex in the mouse.

The great advantage of linking libraries constructed with the use of *Not*I, *Bss*HII, or *Sac*II which have CpG dinucleotides in their recognition sites, is that the clones frequently correspond to HTF islands and hence sites adjacent to genes. Thus, this approach gives a rapid initial screen for potential candidate genes.

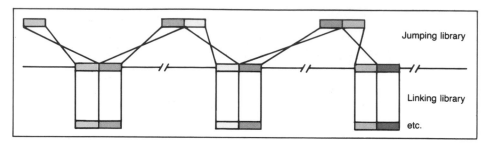

Figure 3.13. Jumping using a linking and rare-cutter jumping library.

8.4 Sequence conservation—'zoo' blots

A common strategy for the identification of genes, particularly if they are not associated with HTF islands, is to search for sequence conservation among a number of different species. This approach was successful for the identification of the Duchenne muscular dystrophy gene. Monaco and colleagues had identified a sequence (PERT87) which was deleted in 10 per cent of patients. They used chromosome walking from PERT87 to screen a Southern blot of genomic DNA from several species ('zoo' blot) and identified a human sequence which was conserved even in chick. This conserved region was used as a probe to screen muscle cDNA libraries which led to the isolation of the DMD gene (36).

8.5 Yeast artificial chromosomes (YAC)

Cosmids have a cloning limit of approximately 50 kb of DNA which has limited their use for chromosome walking projects. This problem has been solved by the maintenance of large fragments of human DNA as yeast artificial chromosomes (YAC; see *Figure 3.14*). This vector is linear and has an origin of replication (ARS), telomere sequences (TEL), a centromere (CEN), and selectable markers, for example, *URA* (37, 38). The capacity of a yeast artificial chromosome (YAC) can exceed 1 Mb but the stability of very large YACs has not yet been very well documented. However, several human genomic YAC libraries have been constructed and the average size of human inserts is as large as 750 kb. This makes chromosome walking much more feasible, simply because a megabase or more can be covered quite rapidly. DNA that is megabases in length can be isolated in YACs and used to construct a complete physical map. This is called a 'contig'.

In addition to chromosome walking, YACs in an area of a disease locus can be identified by screening YAC libraries with *Alu*-PCR products derived from somatic cell hybrids (see Section 7.4; 21). Alternatively, clones obtained by microdissection of a chromosome band can be used to screen the YAC library directly. Both these strategies have been used to isolate sequences across the fragile X at Xq27.

Once a YAC has been isolated, the potential coding regions can be identified

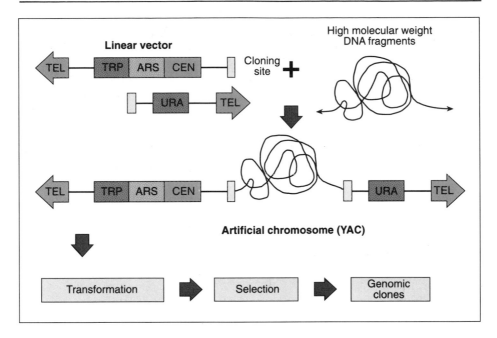

Figure 3.14. Construction of a yeast artificial chromosome (YAC).

by searching for conserved sequences (Section 8.4), by the mapping of HTF islands within the YAC, or by directly screening cDNA libraries with the whole YAC sequence. This last approach was successful for the cloning of the neurofibromatosis gene on human chromosome 17 (39).

YACs are particularly useful for the study of the transcriptional control of genes, since a domain of genomic DNA can be introduced into a cell for study (see ref. 37 for a review of the uses of YACs).

8.6 Exon trapping

Exon trapping is a technique whereby genomic fragments are cloned into a vector with the correct donor and acceptor splice signals such that if the genomic sequence contains an exon, it will be spliced out when vector plus the genomic insert are transiently expressed (*Figure 3.15*). The trapped exon can be identified by PCR amplification of the cDNA derived from the mRNA (40). This technique is particularly important for the identification of disease genes that are not characterized by HTF islands and where the level of expression is very low.

8.7 In situ *hybridization*

In some studies of human genetic disease, a candidate gene is found by positional cloning but confirmation of its role in the phenotype is lacking. In such

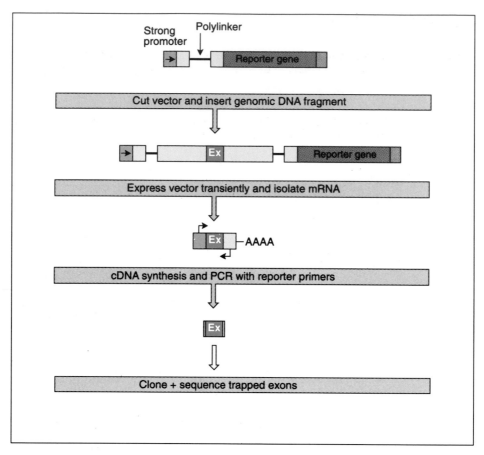

Figure 3.15. Principle of exon trapping.

cases, the tissue distribution of expression of the gene is important as well as its expression pattern during development. A candidate gene can be readily investigated using the latest techniques of *in situ* hybridization to tissue sections. An excellent example of the application of this strategy was in the characterization of a candidate gene for Wilm's tumour, an embryonic kidney tumour thought to arise through aberrant mesenchymal cell differentiation. A gene had been identified (41) which mapped to the region of 11p, often deleted in Wilm's tumour patients. The role of this gene in normal development and tumorigenesis was elucidated by performing *in situ* hybridization with the mRNA on sections of human embryos and Wilm's tumours (42). The gene was expressed in exactly the manner expected from the observed phenotype, and clearly has a specific role in kidney development and a wider role in mesenchymal–epithelial transitions.

9. Further reading

Williams,J.G. and Patient,R.K. (1988) *Genetic engineering, in focus*. IRL Press, Oxford.
Gelehrter,T.M. and Collins,F.S. (1990) *Principles of medical genetics*. Williams & Wilkins, Baltimore.
Glover,D.M. (ed.) (1989) *Cloning—Volumes I, II, and III. A Practical Approach*. IRL Press, Oxford.
Davies,K.E. and Tilghman,S.M. (ed.) (1990) *Genome analysis—Volume I, Genetic and physical mapping*. Cold Spring Harbor Laboratory, New York.

10. References

1. Williams,J.G. (1981) In *Genetic engineering 1*, (ed. R.Williamson), p.1. Academic Press, New York.
2. Williams,J.G. and Patient,R.K. (1988) *Genetic engineering, in focus*. IRL Press, Oxford.
3. Hedrick,S.M., Cohen,D.I., Nielsen,E.A., and Davis,M.M. (1984) *Nature*, **308**, 149.
4. Royer-Pokora,B., Kunkel,L.M., Monaco,A.P., Goff,S.C., Newburger,P.E., Baehner,R.L., Cole,F.S., Curnutte,J.T., and Orkin,S.H. (1986) *Nature*, **322**, 32.
5. Segal,A.W., Cross,A.R., Garcia,R.C., Borregaard,N., Valerius,N.H., Soothill,J.F., and Jones,O.T.C. (1983) *N. Eng. J. Med.*, **308**, 245.
6. Francke,U., Ochs,H.D., de Martinville,B., Giacalone,J., Lindgren,V., Disteche,C., Pagon,R.A., Hofker,M.H., van Ommen, G.J.B., Pearson,P.L., and Wedgwood,R.J. (1985) *Am. J. Hum. Genet.*, **37**, 250.
7. Francke,U. (1984) *Cytogenet. Cell. Genet.*, **38**, 298.
8. Baehner,R.L., Kunkel,L.M., Monaco,A.P., Haines,J.L., Conneally,P.M., Palmer,C., Heerema,N. and Orkin,S.H. (1986) *Proc. Natl. Acad. Sci. USA*, **83**, 3398.
9. Dinauer,M.C., Orkin,S.H., Brown,R., Jesaitis,A.J., and Parkos,C.A. (1987) *Nature*, **327**, 717.
10. Teahan,C., Rowe,P., Parker,P., Totty,N., and Segal,A.W. (1987) *Nature*, **327**, 720.
11. Jolly,D.J., Esty,A.C., Bernard,H.U., and Friedman,T. (1982) *Proc. Natl. Acad. Sci. USA*, **79**, 5038.
12. Thein,S.L. and Wallace,R.B. (1986) In *Human genetic disease, a practical approach*, (ed. K.E.Davies), p.33. IRL Press, Oxford.
13. Ray,P.N., Belfall,B., Duff,C., Logan,C., Kean,V., Thompson,W.W., Sylvester,J.E., Gorski,J.L., Schmickel,R.D., and Worton,R.G. (1985) *Nature*, **318**, 672.
14. Kunkel,L.M., Monaco,A.P., Middlesworth,W., Ochs,H.D., and Latt,S.. (1985) *Proc. Natl. Acad. Sci, USA*, **82**, 4778.
15. Porteous,D.J. (1987) *Trends Genet.*, **3**, 177.
16. Cox,D.R., Kobori,J., Uglum,E., Casher,D., Price,E.R., and Myers,R.M. (1987) *Am. J. Hum. Genet.*, **41**, A162.
17. Pritchard,C. and Goodfellow,P.N. (1986) *EMBO J.*, **5**, 979.
18. Porteous,D.J., Morten,J.E.M., Cranston,G., Fletcher,J.M., Mitchell,A., van Heyningen,V., Fantes,J.A., Boyd,P.A., and Hastie,N.D. (1986) *Mol. Cell. Biol.*, **6**, 2223.
19. Nelson,D.L., Ledbetter,S., Corbo,L., Victoria,M.F., Ramirez-Solis,R., Webster,T.D., Ledbetter,D.H., and Caskey, C.T. (1989) *Proc. Natl. Acad. Sci. USA*, **86**, 6686.
20. Lichter,P., Cremer,T., Borden,J., Manuelidis,L., and Ward,D.C. (1988) *Hum. Genet.*, **80**, 224.
21. Lehrach,H., Drmanac,R., Hoheisel,J., Larin,Z., Lennon,G., Monaco,A.P., Nizetic,D., Zehetner,G., and Poustka,A.-M. (1991) In *Genome analysis, volume I*, (ed. K.E.Davies and S.M.Tilghman), p.39. Cold Spring Harbor Laboratory New York.

22. Ludecke,H.-J., Senger,G., Claussen,U., and Horsthemke,B. (1989) *Nature*, **338**, 348.
23. Hirst,M.C., Rack,K., Nakahori,Y., Roche,A., Bell,M.V., Flynn,G., Christadoulou,Z., MacKinnon,R.N., Francis,M., Littler,A.J., Anand,R., Poustka,A.-M., Lehrach,H., Schlessinger,D., D'Urso,M., Buckle,V.J., and Davies,K.E. (1991) *Nucl. Acid. Res.*, **19**, 3283.
24. Schwartz,D.A. and Cantor,C.R. (1984) *Cell*, **37**, 67.
25. Anand,R. (1986) *Trends Genet.*, **2**, 278.
26. Chu,G., Vollrath,D., and Davis,R.W. (1986) *Science*, **234**, 1582.
27. Erlich,M. and Wang,R.Y.-H. (1981) *Science*, **212**, 1350.
28. Barlow,D.P., and Lehrach,H. (1987) *Trends Genet.*, **3**, 167.
29. Van Dunnen,J.T., Bakker,E., Breteler,E.G.K., Pearson,P.L., and van Ommen,G.J.B. (1987) *Nature*, **329**, 640.
30. Bird,A.P. (1987) *Trends Genet.*, **3**, 342.
31. Lindsay,S. and Bird,A.P. (1987) *Nature*, **326**, 336–8.
32. Estivill,X., Farrall,M., Scambler,P.J., Bell,G.M., Hawley,K.M.F., Lench,N.J., Bates,G.P., Kruyer,H.C., Frederick,P.A., Stanier,P., Watson,E.K., Williamson,R., and Wainwright,B.J. (1987) *Nature*, **326**, 840.
33. Collins,F.S. and Weismann,S.M. (1984) *Proc. Natl. Acad. Sci. USA*, **81**, 6812.
34. Poustka,A. and Lehrach,H. (1986) *Trends Genet.*, **2**, 174.
35. Poustka,A., Pohl,T.M., Barlow,D.P., Frischauf,A.-M., and Lehrach,H. (1987) *Nature*, **325**, 353.
36. Monaco,A.P., Neve,R.L., Colletti-Feener,C., Bertelson,C.J., Kurnit,D.M., and Kunkel,L.M. (1986) *Nature*, **323**, 646.
37. Hieter,P., Connelly,C., Shero,J., McCornick,M.-K., Antonarakis,S., Pavan,W., and Reeves,R. (1990) In *Genome analysis, Volume I*, (ed. K.E.Davies and S.M.Tilghman). Cold Spring Harbor Laboratory, New York.
38. Burke,D.T., Carle,G.F., and Olson,M.V. (1987) *Science*, **236**, 806.
39. Wallace,M., Marchuk,D., Andersen,L., Lechter,R., Odeh,H., Saulino,A., Fountain,J., Brereton,A., Nicholson,J., Mitchell,A., Brownstein,B., and Collins,F.C. (1990) *Science*, **249**, 181.
40. Auch,D. and Reth,M., (1991) *Nucl. Acid. Res.*, **18**, 6743.
41. Call,K.M., Galser,T., Ito,C.Y., Buckler,A.J., Pelletier,J., Haber,D.A., Rose,E.A., Kral,A., Yeger,H., Lewis,W.H., Jones,C., and Housman,D.E. (1990) *Cell*, **60**, 509.
42. Pritchard-Jones,K., Felming,S., Davidson,D., Bickmore,W., Porteous,D., Gosden,C., Bard,J., Buckler,A., Pelletier,J., Housman,D., van Heyningen,V., and Hastie,N. (1990) *Nature*, **346**, 194.

The molecular basis of human inherited disease

1. Introduction

There are several stages from transcription of a gene through to its translation into protein where the cellular machinery can malfunction and result in the diseased phenotype. The most extensively studied group of genetic disorders that illustrate this are the alpha- and beta-thalassaemias. The structure of the α- and β-globin gene clusters on chromosomes 16 and 11 respectively are presented in *Figure 4.1* which include the embryonic, fetal, and adult globin gene arrangements.

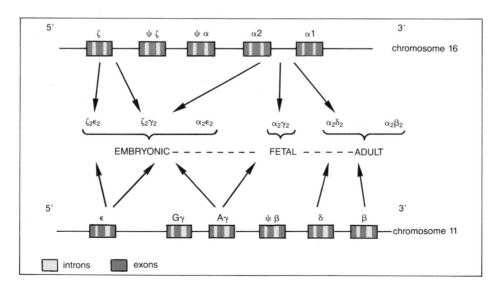

Figure 4.1. Arrangement of human globin genes.

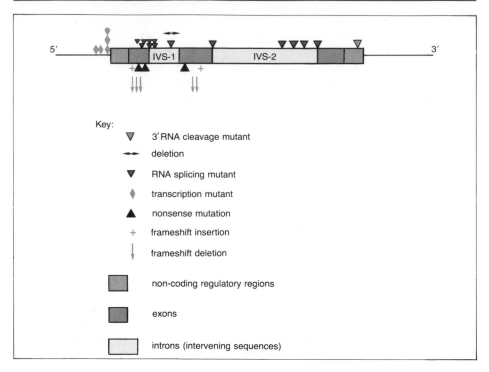

Figure 4.2. β-thalassaemia mutations.

A summary of the types of mutation found within the β-globin cluster are given in *Figure 4.2*, which shows the extent of the heterogeneity for this region (for review see ref. 1). When no β-globin gene product is produced, a β^0-thalassaemia is present, whereas if a small amount of gene product is made, the defect is of the β^+ variety. The β-globin gene cluster has been studied intensively because the genes are expressed in a tissue specific manner and are differentially expressed at different times during development. The DNA sequences involved in this complex control of transcription are not reviewed here except in the context where mutations in these control elements lead to a clinical phenotype. The reader is referred to a recent review by Dillon *et al.* (2).

As other disease loci are being identified, heterogeneous patterns of mutations similar to those found for β-globin are emerging. The underlying mechanism of mutations in some genetic diseases, however, shows some unusual features not seen in the β-globin cluster. These are detailed in Section 3.

2. Mutations affecting the β-globin gene

2.1 Gene deletion

Only one form of β-thalassaemia has been found to be due to a major gene deletion. In this form, which is found in Asian-Indian populations, 619 nucleotides at the 3′ end of the β-globin gene is missing (3). The deletion starts in the large intron (IVS-2) and extends downstream beyond the end of the coding region of the β-globin gene. Other small gene deletions have been identified, but many of these are too small to be detected by routine restriction mapping. Gene deletions are a more common feature for other genes such as the loci for Duchenne muscular dystrophy and the LDL receptor (see Sections 3.3 and 3.2 respectively).

Complex β-thalassaemias where the synthesis of fetal or embryonic β-globin genes is affected (e.g. δβ-thalassaemia and γδβ-thalassaemia) are often associated with quite extensive deletions of DNA (*Figure 4.3;* ref. 1). Some of these are characterized by the persistence of expression of the fetal globin genes into adult life (HPFH syndromes). Typical HPFH is a result of a deletion of both the β- and δ-globin genes. The expression of both γ-globin genes is at a high level, but not sufficiently high to compensate for the complete absence of β-globin production.

The reason for the observed high expression of the γ-globin genes is not fully understood, although it has been proposed that the deletions alter the chromatin structure and thereby modify the gene expression in the region. Alternatively, DNA brought into the region as a consequence of the deletion could contain enhancer activity.

All genes in the cluster are inactive (or deleted) in the rare γδβ-thalassaemia. In one case, β-globin gene expression is absent even though the deletion stops nearly 2.5 kb upstream from the β-globin gene. The cloned β-globin gene from these patients can be expressed normally, suggesting that the DNA sequence introduced into the vicinity of the β-gene by the deletion shuts off β-globin expression in some way, perhaps in a similar fashion to the silencer sequences found associated with other genes (4).

2.2 Transcriptional mutants

Studies in several gene systems and comparison of the DNA sequences flanking normal genes have defined various non-coding sequences which lie 5′, that is

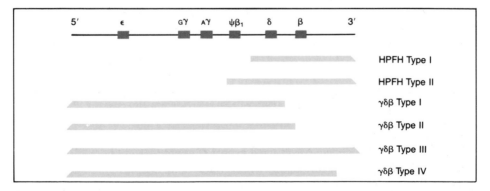

Figure 4.3. Deletions in the β-globin gene cluster.

upstream, of the gene and are essential for normal levels of transcription. Sequences between -75 and -90 base pairs (with respect to the position of the cap site or start of RNA transcription) have the consensus sequence CCAAT and constitute a true promoter element. A mutant involving a C to G change at position -87 leads to a 10-fold reduction in the level of β-globin mRNA produced when this gene is introduced into HeLa cells. This mutation is found in Mediterranean β^+thalassaemia (5).

Another important element in the control of β-globin gene expression is further upstream and has the consensus sequence PuCPuCCC (Pu = purine). The sequence change ACACCC to ACACGC reduces the level of β-globin gene expression $3-5$-fold when compared to normal β-globin gene expression.

A very important 5' control region of eukaryotic genes is the ATA or TAT box found at about 30 bp upstream of the first AUG codon. Two transcriptional mutants involving changes at position -28 have been reported. In a Chinese form of β^+-thalassaemia, the normal sequence, CATAAAA, becomes CATGAAA and in a Kurdish patient, the normal sequence has been found to be mutated to CATAGAA. In both cases, the level of mRNA falls to about $20-30$ per cent that of the normal and the mutation has only mild effects (6,7).

2.3 RNA processing mutations

Intron or intervening sequences (IVSs) are a common feature of most eukaryotic genes and they are spliced out during normal processing of the precursor mRNA (see ref. 8 for review). Consensus sequences for splice junctions have been derived from the analysis of different genes: the 5' donor sequence has an absolute requirement for GT and the 3' acceptor sequence for AG. Mutations in the β-globin gene either alter these sequences directly or introduce other cryptic donor or acceptor sites.

Two β-thalassaemia genes have been identified in which the G of the GT at the donor site is mutated to A at either IVS-1 or IVS-2 (9). This leads to alternative processing of the mRNA at other donor-like sequences located in either the coding exon or intervening (intron) sequences.

Next to the invariant dinucleotides at the junction between the coding regions and the introns, there are preferred sequences that vary among different genes. A second class of processing mutant includes nucleotide changes within these regions. Mutations result in the abolition or reduction of normal splicing at the donor site and a low level of alternative RNA processing (see *Figure 4.4;* ref. 5).

Nucleotide substitutions within an intervening sequence (intron) may generate new splicing signals (see *Figure 4.5;* ref. 10). The severity of the resulting phenotype will depend on the fraction of the transcripts processed abnormally, compared to the normally spliced mRNA.

Obviously, nucleotide substitutions which occur within the coding regions of the β-globin gene may also introduce a cryptic splicing site. The β-globin chain variant of HbE is such a mutation, where there is an amino acid replacement (Lys to Glu) at position 26 of the β-polypeptide (11). The β^E chain is under-produced in erythroid cells and its mRNA is deficient in relation to the total

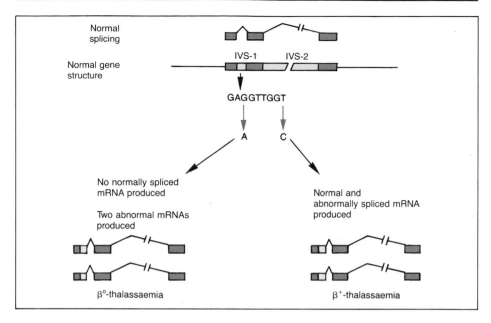

Figure 4.4. Splice junction mutations in the β-globin gene.

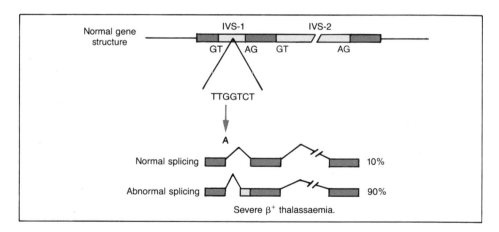

Figure 4.5. Production of a new splice site in the β-globin gene.

β-globin mRNA. The region of exon 1 which contains codons 24 – 27 is similar in sequence to the donor splice site (*Figure 4.6*). The G to A substitution activates the cryptic splice site at the junction of exon 1 and IVS-1, which is then utilized at a low level. The competition of this new exon 1 donor site with the authentic IVS-1 donor site results in delay as well as alternative β-globin mRNA processing. This mutant demonstrates the complexity of RNA processing.

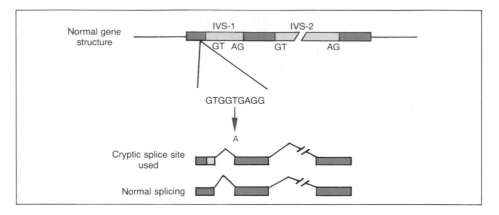

Figure 4.6. Processing of β^E-globin RNA transcripts.

2.4 RNA cleavage signal mutations

A consensus sequence AAUAAA found to occur at the 3′ end of eukaryotic mRNAs is thought to be important in cleavage and polyadenylation. A mutation in this consensus sequence has been found to be associated with a mild form of β-thalassaemia (12). The sequence AATAAA is substituted by AACAAA. The RNA synthesized from affected individuals is processed at the next 3′ AATAAA, which occurs a further 90 bp downstream.

2.5 Nonsense and frameshift mutations

Nonsense mutations are alterations in the sequence such that no functional protein is produced due to the premature termination of translation. A single base mutation introduces an in-frame termination at codons 17 or 39 of the β-globin gene (13,14).

Other mutations that cause the production of non-functional mRNA are frameshifts. Single base insertions, or deletions alter the reading frame of the protein. There are several examples of frameshifts resulting from deletions within the β-globin gene. They are thought to occur during DNA replication and are possibly due to slippage in regions where short direct repeats occur. This hypothesis has been proposed because there appears to be a hot spot for such mutations involving codons 6–8 in the β-globin gene.

In many of the instances described above, the non-functional mRNAs are found only at very low levels in erythroid cells. This may reflect the intrinsic instability of the mutant mRNA, but the reasons for this are not fully known. It is of interest that in Haemoglobin Constant Spring, the α-globin chain has 172 residues instead of 141 because of a mutation from UAA (stop) to CAA (Gln). Translation of part of what is normally a non-coding region somehow renders the altered mRNA susceptible to the action of nucleases (see ref. 15 for review of α-globin chain termination mutants).

3. Mutations in other genetic disorders

The advent of rapid methods for mapping chromosomal regions has led to the identification of many disease mutations. Some of these have been characterized in detail and some general mechanisms for deletions, translocations and duplications are emerging from sequencing studies (see review, ref. 16). The examples given below were chosen because each one displays a particularly novel type of mutation underlying human genetic disease.

3.1 α-thalassaemia

As shown in *Figure 4.1* at the beginning of this chapter, the α-globin genes are present as duplicate copies on chromosome 16. This arrangement makes them prone to deletion and duplication due to unequal crossing over during meiosis, as illustrated in *Figure 4.7*. The two chromosomes misalign, resulting in one chromosome 16 with three copies and the other with just a single copy of the gene. Haplotypes with four α-globin genes on one chromosome have also been observed.

More recently, it has been demonstrated that homologous recombination within the α-globin cluster can also occur between *Alu* repeats, which results in the generation of deleted chromosomes (17). The *Alu* repeats are in the same orientation on the chromosome and therefore align by interchromosomal pairing.

An intriguing class of deletions has been found in α-thalassaemic patients. Many of these are a consequence of the fact that the α-globin cluster lies within a few hundred kilobases of the telomere of chromosome 16p. For example, some α-thalassaemia patients have intact α-globin genes but possess a truncated chromosome 16, where the truncation event has removed elements upstream from the genes which are important for their expression (18). In other cases, the telomeric deletion of the chromosome extends to sequences proximal to the α-globin genes (19). These patients are mentally retarded as well as suffering

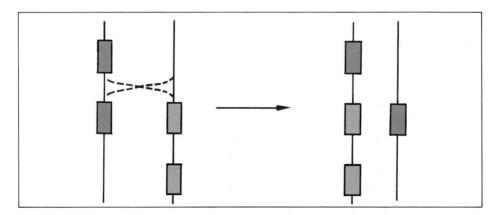

Figure 4.7. Unequal crossing-over and deletions in the α-globin genes.

from α-thalassaemia. This mechanism of chromosome truncation may be quite a common one in human genetic disease. For example, Wolf – Hirschhorn and Miller – Dieker syndromes are associated with deletions of chromosomes 4p and 17p respectively.

Cases of mental retardation associated with α-thalassaemia have also been described where telomeric exchange is involved. A cytogenetically invisible sub-telomeric translocation between chromosomes 1 and 16 has been described in a family with mental handicap and thalassaemia (20). The existence of such telomeric translocations has also been demonstrated with *cri du chat* syndrome on chromosome 5 (21). Such telomeric exchanges may underlie other genetic syndromes as yet unrecognized.

There are some families segregating α-thalassaemia with mental retardation where the affected individuals are male and the α-globin cluster appears intact (22,23). Affected individuals in these families are more severely mentally retarded compared with the deletion/translocation cases described above and have dysmorphic features. These families segregate this syndrome in a manner consistent with an abnormality in a trans-acting factor necessary for the normal regulation of the α-globin complex. In view of the fact that all of the patients so far described are male, this factor might well be localized on the X-chromosome.

3.2 Familial hypercholesterolaemia

About 10 per cent of patients who suffer from hypercholesterolaemia have defects in their low density lipoprotein (LDL) receptor gene which are the result of a gene deletion or duplication. The explanation for this high rate is the presence of *Alu* repeats in various orientations along the gene (24). Thus, gene deletions or duplications occur by homologous recombination between chromosomes at meiosis in a similar way to those observed at the α-globin gene locus.

More recently, an LDL receptor gene deletion has been postulated to occur by intrachromosomal pairing (*Figure 4.8;* ref. 25). This again is due to the misalignment of sequences, but here it is thought that the chromosome loops out to allow the pairing of *Alu* repeats organized in opposite orientations relative to each other.

Hypercholesterolaemia has been shown to be due to either deletions or duplications of parts or all of the gene sequence of the LDL receptor (26).

3.3 Duchenne muscular dystrophy

Duchenne muscular dystrophy (DMD) is the most common X-linked recessive disease in man, affecting 1 in 3000 males. Affected boys are usually wheelchair-bound by the age of 12 years and dead by the age of twenty. In the milder form of the disorder, Becker muscular dystrophy (BMD), affected individuals do not need to use a wheelchair until much later in life and they can live a normal life span. The first clue to the localization of the DMD gene came from the observation of females affected by the disease with balanced X;autosome translocations (27). The autosomal breakpoints were different in each case, but

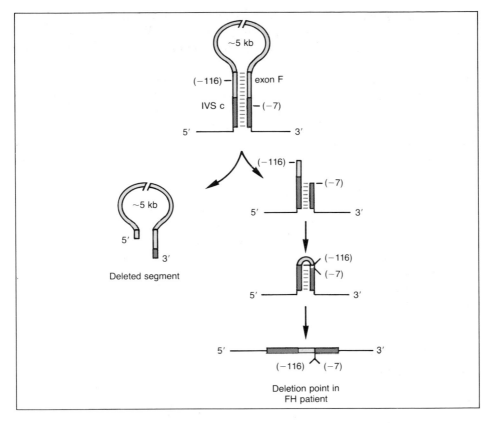

Figure 4.8. Potential mechanism for gene deletion by pairing of *Alu* sequences in the LDL receptor gene. FH denotes familial hypercholesterolaemia.

the X-chromosome breakpoint lay within Xp21 (28). In balanced X;autosome translocations, it is the translocated X-chromosome that remains active and the normal X-chromosome which is inactive. Thus, if the breakpoint in Xp21 disrupts gene expression, the females will suffer from the disease like male hemizygotes. This localization of the mutation to Xp21 was confirmed by linkage studies using RFLPs flanking the Xp21 band (29). The RFLPs segregated in the same way in DMD families as in BMD families (30). This suggested that these two phenotypes might be mutations in the same gene sequence.

Kunkel and his colleagues (31) isolated clones from within the DMD locus by genomic enrichment using the DNA from a patient with a deletion within Xp21, suffering from DMD, CGD, retinitis pigmentosa, and McLeod phenotype. This strategy is described in Section 7.2 of Chapter 3. Other clones were identified by the cloning of the X;21 translocation breakpoint in a female suffering from the disease (see Section 6 of Chapter 3; ref. 32). The first section of the coding sequence was identified by Kunkel's group by screening chromosome walks of their original clones against 'zoo' blots for sequence conservation (33). Subsequently, the 14 kb mRNA has been shown to be expressed in human fetal and adult muscle, and at lower levels in brain tissue (34,35).

Pulsed field gel electrophoresis has shown that the 14 kb DMD mRNA covers a genomic distance of at least 2.3 Mb making this the largest gene complex so far reported (36, 37, 38). In comparison, the Factor VIII mRNA of 9 kb covers 200 kb of genomic sequence. The average size of an exon is 150 bp and there are over 70 exons in the whole DMD gene. The large size of the gene in part explains the high mutation rate in this disease. However, there is a high frequency of deletion within the DMD locus where 70 per cent of patients possess a deletion (34, 39, 40). Therefore, there may be sequences within Xp21 that promote such deletions.

Fifty per cent of both DMD and Becker muscular dystrophy (BMD) patients possess deletions near the centre of the gene extending towards the 3' end. The 5' endpoints of these deletions are spread across one of the large introns of the gene (41). The underlying cause of the high deletion rate in this portion of the gene cannot be explained on the basis of unequal meiotic crossing over, as a corresponding high frequency of duplications has not been observed for the same region. Although duplications have been observed, they are seen in less than 5 per cent of the cases. Sequencing of the deletion breakpoints in different patients has revealed a common 6 bp motif but no obvious mechanism for the deletions has been determined (42).

A second region which is frequently deleted in patients lies near the 5' end of the gene which also contains a large intron. For diagnostic purposes, advantage has been taken of the favoured sites of deletion in designing multiplex PCR primer sets which can detect 95 per cent of deletions in the gene (43). Carrier detection of the disease is still problematical, since gene dosage studies with the cDNA can be difficult to interpret. More recently, fluorescence *in situ* hybridization has been used with cosmids from the gene regions most often deleted (44). Carrier status can be unequivocally determined by the hybridization with one X-chromosome but not the other. Diagnosis of mutations has recently been carried out by investigating the ectopic expression of dystrophin. Very low levels of expression of the gene in lymphocytes can be detected by PCR (45) and thus the mutant mRNA can be investigated. This type of analysis can be extended to carriers where both mutant and wild type PCR products are seen.

Antisera raised against expressed portions of the DMD gene sequence identify a 427 000 kd protein in adult and fetal muscle. The molecule has been called 'dystrophin' because its absence causes dystrophy (46). Studies of the expression of the gene and the protein product show that many DMD patients produce very little or no dystrophin in their muscle cells, whereas the much milder BMD patients produce truncated products (47). This suggests that the milder BMD is due to the synthesis of lower molecular weight protein products which still retain some function in muscle. Thus, deletions in BMD patients must leave the protein in phase whereas the deletions in DMD patients presumably destroy the open reading frame of the protein, remove portions vital to the function of dystrophin, or prevent transcription (48).

Although 70 per cent of DMD patients have deletions in the dystrophin gene, mutations in the remaining 30 per cent of patients have not as yet been well defined. Some of these are likely to be point mutations similar to those described at the β-globin locus. Since the majority of DMD patients show little or no dystrophin protein production, the sequence changes probably affect transcription either by the introduction of stop codons or perhaps by the modification of the promoter elements of the gene.

Dystrophin is estimated to account for approximately 0.002 per cent of the total protein in a muscle cell which explains why conventional biochemical techniques were not able to detect its absence in patients. The molecule has domains which show homology to α-actinin and spectrin. It is localized to the cytoplasmic face of the muscle membrane, but its exact function remains to be elucidated (for a review see ref. 49).

The very large size of the DMD gene product and its tissue specificity make replacement therapy very difficult; however, analysis of the function of the gene should lead to better treatments. One fruitful strategy is the study of animal models of DMD. Similar genetic defects in both mice and dogs have been reported. The *mdx* mutation in the mouse is a point mutation in the murine equivalent of the human DMD gene, although the mouse is affected much more mildly than its human counterpart (50). Molecular studies show that the *mdx* mouse has reduced levels of dystrophin mRNA in muscle and brain. Correction of the defect in the mouse has been approached by creating an *mdx* mouse transgenic for the full length dystrophin cDNA (51) and by direct injection of the cDNA into muscle (52). In both strategies, dystrophin positive muscle fibres were found. Such experiments should lead to the elucidation of the functionally important domains of the protein as well as opening possible avenues for future gene therapy (see Section 6). In the dog model, no dystrophin is produced and the muscle cells are affected in much the same way as in man (53).

The isolation and study of the dystrophin gene is summarized in *Figure 4.9*.

3.4 Haemophilia A

Haemophilia A is caused by mutations in the Factor VIII gene which encodes a protein involved in blood clotting. The gene is spread over 220 kb of genomic DNA sequence and the processed mRNA is 9 kb in length (54). Consistent with this size, 10 per cent of patients are the result of deletions of part or all of the gene (55). One of the common point mutations in the Factor VIII gene is that which results in a C to T transition (56). In certain parts of the gene this sequence substitution introduces a stop codon into the protein. The mutation can often be readily identified because it also removes a cutting site for the restriction enzyme *Taq*I. *Taq*I recognizes the sequence TCGA and the common mechanism for the C to T transition is deamination of the methylated cytosine which occurs adjacent to the G residue to give TTGA. This deamination of methylcytosine

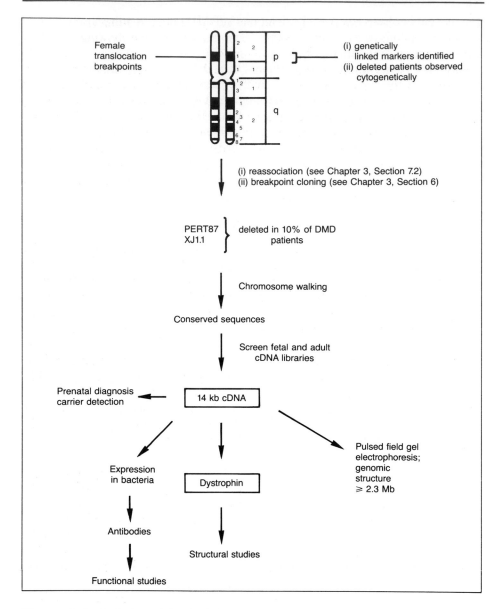

Figure 4.9. Isolation of the DMD gene.

in eukaryotic DNA is a well documented phenomenon and it is also the reason why *Taq*I is so efficient in revealing restriction fragment length polymorphisms.

One intriguing mutation at the Factor VIII locus is an insertion of an L1 sequence (57). The L1 sequence belongs to a family of long, interspersed, repetitive elements present in approximately 10^5 copies throughout the genome.

L1 elements possess an A-rich 3' end and two long open reading frames (*orf*-1 and *orf*-2), the second of which encodes a potential polypeptide showing homology with reverse transcriptases. This structure suggests that L1 sequences might function as non-viral retrotransposons. The L1 sequence was found to be integrated into an exon of the Factor VIII gene in two unrelated patients. Both insertions contained the 3' portion of the L1 sequence, including the poly(A) tract, and created target duplications of at least 12 nucleotides of the Factor VIII gene. Their 3' trailer sequences following *orf*-2 were nearly identical to the consensus sequence of L1 cDNAs. Thus, the insertion of L1 sequences, involving retrotransposition of DNA sequences through an RNA intermediate into a new location in the genome, may represent a not uncommon mechanism for the creation of new mutations in human disease.

3.5 Cystic fibrosis

Cystic fibrosis (CF) is the most common autosomal recessive disease in Caucasian populations with a carrier frequency of 1 in 20 people. With the advent of better antibiotics, better nutrition, and improved chest physiotherapy, the life span of CF patients has increased, but most still die in their twenties. The mutation has been localized to chromosome 7q by linkage analysis; the gene was hard to find because it was not marked by deletions or translocations in patients. The position of the gene was narrowed down to 1600 kb of DNA by linkage analysis. Further refinement of the gene location was then obtained by analysing linkage disequilibrium between new markers and the disease. The region defined by the genetic map was investigated using chromosome jumping techniques and saturation cloning. Eventually, the gene was identified by screening a conserved fragment of DNA against a sweat gland cDNA library (58,59). The gene is 250 kb in length and encodes a mRNA of 6.5 kb corresponding to a protein of 1480 amino acids (for reviews, see refs 60–62).

The CF protein is known as the cystic fibrosis transmembrane conductance regulator (CFTR). About 70 per cent of patients in Britain and North America have the same single amino acid deletion (phenylalanine) at position 508 (ΔF508). This occurs in the first fold of the protein and is likely to have a significant effect on the conformation of the protein, and hence its function. The ΔF508 mutation is less frequent in other European countries, and more than 100 other mutations have been described, most of which are uncommon. In view of the high carrier frequency of CF, pilot studies of population screening for the defect have been initiated in some countries such as Britain. The ethical issues are being carefully analyzed since such a procedure with only 70 per cent of CF carriers carrying the ΔF508 mutation could cause much anxiety. Couples where only one partner has the ΔF508 mutation have their risk of having an affected child increased from 1 in 2500 without the test to 1 in 400 with the test, which may not be very helpful.

Although the exact role of CFTR in regulating the transport of chloride ions out of epithelial cells is not understood, recent experiments provide evidence that it may be a specific type of chloride channel (for review see ref. 63).

However, in sequence and domain structure, the protein seems to be a typical member of the ATP-binding cassette (ABC) superfamily of transporters. Thus, CFTR may well turn out to be multifunctional, acting as both a chloride channel and a transporter (for an unidentified substrate).

Although CF is a multi-organ disease, the lungs are a critical target for treatment. Aerosols may be one way of delivering CFTR (wrapped in a lipid bilayer, for example) but enabling CFTR to then be taken up into the membrane of the right cells will be a major challenge. Although there are problems for pharmaceutical approaches or gene therapy for CF, the elucidation of the function of CFTR means that clinicians can look forward to a better and more effective treatment of CF in the future.

3.6 Neurofibromatosis type 1

Von Recklinghausen neurofibromatosis type 1 (NF1) is an autosomal dominant disorder with an incidence of 1 in 3000 individuals in all ethnic groups. It is characterized by patches of skin pigmentation and neurofibromas. New mutations are frequent, suggesting that, like DMD, rearrangements of genomic DNA in the vicinity of the NF1 locus may lead to the phenotype. The gene was cloned by taking advantage of the occurrence of two translocations involving chromosome 17q in two unrelated families segregating for the disease. The locus is of particular interest because it encodes four genes, only one of which appears to be mutated in NF1 patients (64 – 66).

The NF1 gene is ubiquitously expressed 13 kb mRNA and studies of mutations suggest that the disease is caused by the loss of function of the NF1 gene product. The observations are consistent with the proposal that the NF1 gene is a tumour suppressor and the mechanism of action is recessive at the cellular level. The NF1 gene shows functional homology to mammalian GAP and yeast IRA proteins which interact with RAS proteins (67). RAS proteins are known to affect cellular proliferation and differentiation but whether the NF1 gene regulates or mediates the RAS function remains to be determined. It is also not yet clear why the disease appears so prominently in neural crest derivatives when the gene is ubiquitously expressed.

The other three transcripts at this locus lie within the first major intron of the NF1 gene and are transcribed in the opposing orientation. It may be that mutations at the NF1 locus also affect their expression and *vice versa*. Two of the gene products, EV12 (an oncogene) and a protein designated RC1, appear to be transmembrane proteins without specified function. The other gene, OMGP, is an externally located cell adhesion molecule potentially active in mediating proper cell motility and differentiation in brain development. EV12 is also expressed in brain. It is possible that expression of both EV12 and OMGP are modified in NF1 patients who have learning disabilities. The higher frequency of juvenile chronic myelogenous leukaemias found among NF1 patients could possibly be explained by the disruption of expression of either EV12 or RC1, or both.

3.7 Genomic imprinting: Prader – Willi syndrome

The concept of genomic imprinting was introduced to explain the observation that genetic information from both maternal and paternal genomes is necessary for the normal development of an embryo. In the parental germ lines, the maternal and paternal copies are marked differently such that they are functionally non-equivalent in the embryo (for reviews see refs 68 – 71). This phenomenon is now widely recognized in mammals, although the underlying mechanisms involved have not yet been elucidated. It is probably responsible for irregular patterns of inheritance and variable expression of a number of human genetic disorders. The role of imprinting is clearly demonstrated in the expression of the Prader – Willi syndrome and it has also been implicated in the increase of severity of myotonic dystrophy with transmission through several generations.

The clinical features of the Prader – Willi syndrome include hypotonia in infancy, obesity, hypogonadotrophic hypogonadism, and mental retardation. Some patients possess chromosomal deletions of 15q11 – 13, although more than half do not have visible chromosomal abnormalities. DNA studies in patients show that the syndrome is associated with the deletion of the paternal chromosome 15. Angelman's syndrome is also associated with a deletion in a similar region of chromosome 15. The clinical features are quite different from that of Prader – Willi and include unusual and frequent laughter and bizarre repetitive symmetrical ataxic movements. The deletions in Angelman's syndrome typically involve the maternally inherited chromosome 15. Although the deletions in these two syndromes may not be identical, the DNA mapping studies suggest that there is a common overlapping region which may contain genes whose function is determined by genomic imprinting.

In non-deletion cases of Prader – Willi syndrome, it has been shown that both copies of chromosome 15 are inherited from the mother (uniparental disomy; ref. 72). Both isodisomy and heterodisomy have been observed, strongly suggesting that the phenotype is caused by the lack of inheritance of the paternal chromosome 15. Uniparental disomy has also been described in Angelman's syndrome, but it is much rarer than in Prader – Willi syndrome, suggesting that the pathogenic mechanisms may be different.

There are now several human genetic disorders where genomic imprinting has been implicated, including cancer (73). Comparison with the map of the mouse also provides important clues as imprinted regions in the mouse are often imprinted in the homologous regions in man. However, it is not yet known whether individual genes, clusters of genes, or genomic segments are imprinted, so care must be taken in suggesting such homologies.

3.8 The fragile X syndrome

The fragile X syndrome is the commonest inherited cause of mental impairment affecting 1 in 1500 males (for reviews, see ref. 74). Approximately one third of female carriers of the mutation are affected, while 20 per cent of males inheriting the mutant chromosome are phenotypically normal transmitters of the disorder (so-called normal transmitting males, or NTMs) (75, 76). The disease is associated with the expression of a folate-sensitive fragile site at Xq27.3, although the level of expression can vary from a few per cent to as high as 50 per cent of cells (77, 78). Typically, daughters of normal transmitting males are not retarded and either do not express the fragile site or express it only at a low level. This can make genetic counselling particularly difficult.

The unusual inheritance pattern observed in the fragile X syndrome has led to the hypothesis that the development of the full phenotype involves a two-step process (79–81). The first non-phenotypic, or premutation, event is converted into the full mutation only after passing through oogenesis. Several hypotheses have been proposed as to the nature of this second mutation event, including genomic imprinting and amplification of DNA sequences at the fragile site. Recently, evidence has accumulated for the possible role of both of these in the progression to the fragile X phenotype. The DNA of fragile X patients is hypermethylated at a CpG island close to the fragile site (82, 83). These CpG dinucleotides are not generally methylated in normal transmitting males or in their daughters but do become methylated in the affected grandsons, suggesting that genomic imprinting could be occurring via methylation in early oogenesis. More recently, several workers have reported the insertion/amplification of a CGG repeat adjacent to these CpG residues in fragile X patients and normal transmitting males (84–86). This insertion/amplification event can be readily observed on Southern blots by a change of fragment size. Normal individuals have up to about 45 copies (mean number 29) of the CGG repeat whereas carriers of the mutation have more than 54 copies. This amplification results in the increase in size of DNA fragments in Southern blots and is diagnostic of the disease. Normal transmitting males (NTMs) have amplifications of CGG which increase the DNA fragment size adjacent to the CpG island by up to 500 bp compared to the normal pattern. This fragment is rarely altered in size when passed on to the daughters of NTMs. Thus, these phenotypically normal daughters show a doublet on Southern blot analysis. These females pass the mutation on to their daughters, when the fragment can either remain unchanged or increase in size due to the further amplification of the CGG repeats. Affected males often show a smear of fragments indicating somatic variation and the presence of CGG amplification of more than 1 kb. The instability of the CGG repeat sequence appears to be directly proportional to its size. Thus, females with X-chromosomes with larger amplifications are more likely to have affected offspring than females with mutations with smaller amplifications. These observations explain the unusual segregation patterns observed in the fragile X syndrome, the so-called 'Sherman paradox'.

The CGG repeat sequence has been mapped very close to the methylated CpG island and, as explained above, the presence of extra copies of the CGG repeat correlates well with the expression of the phenotype (85). The gene associated with the CpG island encodes a 4.8 kb mRNA expressed in brain and lymphocytes known as FMR-1 (86). The product does not show very significant homology to any previously characterized proteins, although there is some sequence similarity to protamines. The gene product may be a DNA binding protein which plays a role in the determination of chromosome structure (86).

Families have been described where individuals express the fragile site in Xq27 at high levels but where no amplification of the trinucleotide repeat is observed (87). These families might well be segregating an additional rare fragile site in Xq27. It will be interesting to determine the molecular basis of this fragility and its relationship, if any, to the amplification of DNA sequences.

The amplification of a trinucleotide sequence is not a unique mechanism in human genetic disease. X-linked spinal and bulbar muscular atrophy and myotonic dystrophy have recently been found associated with the presence of an amplification of CAG and CTG respectively (88). Again, as the number of repeats increases the sequence becomes more unstable. The amplification of trinucleotide repeats may be a relatively common mechanism for mutation in the human genome.

4. Mouse models of human genetic disease

Studies of human genetic disease are difficult because of the small numbers of large informative families available for linkage studies and the small quantities of tissues that can be extracted for the analysis of candidate gene sequences. Studies of homologous genes in the mouse overcome these problems. The mouse genome is now extensively mapped with highly informative microsatellite markers. The map is powerful enough such that genes involved in polygenic traits can be located. Genes which play a major role in diabetes type I have been mapped in this way (89). If a mouse model for a genetic disease already exists, then this provides a very useful resource for identifying the corresponding human genes. The studies of the *mdx* mouse have been very informative in the study of DMD and the role of dystrophin (90; see Sections 3.3 and 6 in this chapter).

In cases where natural mutations do not exist in the mouse, these may be generated by mutagenesis (91). Insertional mutagenesis has greatly facilitated the molecular genetic analysis of gene expression and involves the generation of transgenic mice after the integration of foreign DNA into the host rodent genome (92). A common approach is to infect embryos with a retrovirus. Alternatively, cloned DNA can be introduced into the pronucleus of fertilized mouse oocytes. Such pronuclear injection has also been very useful for the production of new mouse strains expressing the introduced genetic information in a predictable and tissue-specific manner. This approach has an important role to play in the elucidation of factors which control gene expression.

An example of the application of this technology to study dominant negative mutations is given by the experiments in which a mutant collagen gene known to cause perinatal lethal disease in man was introduced into transgenic mice. As little as 10 per cent mutant RNA of the total pro-alpha (I) collagen RNA caused a dominant perinatal lethal phenotype in the mouse (93).

Transfection and infection of foreign DNA can also be performed with embryo-derived (ES) cells (94). These are continuous cell lines derived from mouse blastocyst stage embryos and can contribute to all cell lineages, including the germline, when microinjected into blastocyst stage embryos. Targeted disruption of endogenous genes in ES cells *in vitro* by homologous recombination has opened many new exciting avenues for the generation of mouse models of human genetic disease and for the assay of gene function (95, 96).

There is no doubt that the development of mouse models for human disease will provide not only a valuable system for understanding disease pathology, but also a means to test gene therapy strategies and potential treatments in the future.

5. The human genome mapping project

Many human genetic disorders are so rare that the corresponding genes cannot be localized by family studies, and for many other diseases there is no genetic basis. The molecular analysis of these clinical phenotypes must await the unravelling of the sequence of the whole human genome, when candidate genes can be investigated. This human genome mapping project is now underway and should provide a wealth of information on the control of human gene expression for both the wild type and mutations, in the future. Currently, a complete genetic and physical map of the human genome is being constructed and methods of rapidly identifying genes are being elucidated. In parallel genomic maps of other organisms (e.g. *Saccharomyces cerevisiae, Arabidopsis thaliana, Drosophila melanogaster, Caenorhabditis elegans,* mouse, pig) are being determined as a starting point for major sequencing projects. Comparative mapping is a powerful method of determining the function of human genes and the mouse may provide many useful models for human genetic disease (see Section 4).

Although the technology is not yet available for rapid and relatively cheap large-scale sequencing, sequencing projects have been initiated for *E. coli, C.elegans* and yeast chromosomes. In the human genome, initial sequencing projects are concentrating on cDNA from complex tissues such as fetal brain. Although the human genome project is likely to take at least 15 years to complete, the benefits to biology of this endeavour should be felt immediately as the information is collected. There is no doubt that the next decade will be an exciting time for molecular medicine.

6. Gene therapy

The ultimate aim of any research into human genetic disease must be cure of the disease by somatic or germ line therapy. For both ethical and practical reasons, most research has been focused on somatic gene therapy. Gene therapy can be carried out by removing defective cells from the patient, introducing the functional gene, and replacing the corrected cells back into the patient (for review, see refs 97,98). Alternatively, the gene product might be targeted directly to the appropriate tissue or organ within the patient. The latter approach is very difficult, but much progress has been made in the former approach using retroviral vectors as a means of introducing genes into cells. For example, immunotherapy of patients with advanced melanoma, using tumour-infiltrating lymphocytes modified by retroviral gene transduction has been reported (99).

Diseases of the bone marrow provide suitable models for the development of gene therapy because the relevant cells are reasonably accessible and stem cells can be infected by retroviruses. Thus, much attention has been given to the correction of disorders of haemoglobin production, such as the thalassaemias, and to the T cell specific disease SCID resulting from deficiency in ADA (adenine deaminase). In these cases, the correct gene is introduced using retroviruses but, to date, the levels of expression of the transduced genes have been low.

In general, the development of effective gene therapy must satisfy several objectives:

(a) There must be an adequate source of the correct gene.

(b) It must be a low frequency or single treatment. The correction of non-dividing cells must be stable or the stem cells must be corrected.

(c) There must be a high efficiency delivery system to the target cell without overflow into other cell types.

(d) There must be regulated growth (selection) of corrected cells.

(e) There must be correct regulation of the introduced gene product.

(f) Gene therapy should be performed before irreversible pathology occurs.

(g) The benefit of the treatment for the individual must outweigh the risk both to the patient and to the community.

There has been much progress in the infection of cells with retroviruses, and more recently adenoviruses and HSV (Herpes simplex virus) have been tried. The advantage of the latter two is that they do not require cell mitosis for transgene expression. Much needs to be learned about the biology of stem cells as well as the viral delivery vectors, but there is little doubt that rapid advances are likely in this area in the next few years as the molecular basis of the more frequent disorders is unravelled.

In the case of DMD, in addition to the introduction of dystrophin into muscle cells with retroviruses, direct injection of functional myoblasts (myoblast transfer) is being investigated (90). However, this approach is still difficult to envisage as a cure because of the number of myoblast cells that would be needed to correct the defect and the problems of rejection. Direct injection of the dystrophin gene into muscle cells of the *mdx* mouse (the mouse model for DMD) has been shown to result in the correct localization of dystrophin at the muscle cell membrane. At present, this work holds little promise of gene therapy for DMD because the levels of dystrophin expression achieved are very low, but they are proving to be very illuminating with regard to the elucidation of the function of dystrophin. Hopefully, these approaches will lead to other strategies for alleviating this devastating disease in the future.

7. Further reading

Beebee,T. and Burke,J. (1992) *Gene structure and transcription, in focus*, (2nd edn). IRL Press, Oxford.
Davies,K.E. (ed.) (1986) *Human genetic disease, a practical approach*. IRL Press, Oxford.
Davies,K.E. (ed.) (1988) *Genome analysis, a practical approach*. IRL Press, Oxford.
Davies,K.E. and Tilghman,S.M. (ed.) (1990) *Genome analysis, Volume I, Genetic and Physical Mapping*. Cold Spring Harbor Laboratory, New York.
Davies,K.E. and Tilghman,S.M. (ed.) (1991) *Genome analysis, Volume II, Gene expression and its control*. Cold Spring Harbor Laboratory, New York.
Davies,K.E. and Tilghman,S.M. (ed.) (1991) *Genome analysis, Volume III, Genes and phenotype*. Cold Spring Laboratory, New York.
Friedmann,T. (ed.) (1991) *Therapy for genetic disease*. Oxford University Press.
Gelehrter,T.D. and Collins,F.C. (1990) *Principles of medical genetics*. Williams & Wilkins, Baltimore.
Weatherall,D.J. (1991) *The new genetics and clinical practice*. Oxford Univesity Press.
Tsui,L.-C., Romeo,G., Greger,R., and Gorini,S. (1991) *The identification of the CF (cystic fibrosis) gene, recent progress and new research strategies. Advances in experimental medicine and biology, Volume 290*. Plenum Press, New York.

8. References

1. Orkin,S.H. and Kazazian,H.H. (1984) *Annu. Rev. Genet.*, **18**, 131.
2. Dillon,N., Talbot,D., Philipsen,S., Hanscombe,O., Fraser,P., Lindenbaum,M., and Grosveld,F. (1991) In *Genome Analysis, Volume II*. (ed. K.E. Davies and S.M.Tilghman). Cold Spring Harbor Laboratory, New York.
3. Orkin,S.H., Old,J.M., Weatherall,D.J., and Nathan,D.G. (1979) *Proc. Natl. Acad. Sci. USA*, **76**, 2400.
4. Kioussis,D., Vanin,E., de Lange,T., Flavell,R.A., and Grosveld,F.G. (1983) *Nature*, **36**, 662.
5. Treisman,R.A., Orkin,S.H., and Maniatis,T. (1983) *Nature*, **302**, 591
6. Orkin,S.H., Sexton,J.P., Cheng,T.-C., Goff,S.C., Giardina,P.J.V., Lee,J.I., and Kazazian,H.H. (1983) *Nucl. Acid. Res.*, **11**, 4727.
7. Poncz,M., Ballantine,M., Solowiejczyk,D., Barak,I., Schwartz,E., and Surrey,S. (1982) *J. Biol. Chem.*, **257**, 5994.
8. Breitbart,R.E., Andreadis,A., and Nadal-Ginard,B (1987) *Ann. Rev. Biochem.*, **56**, 467.
9. Treisman,R.A., Proudfoot,N.J., Shander,M., and Maniatis,T. (1982) *Cell*, **29**, 903.
10. Westaway,D. and Williamson,R. (1981) *Nucl. Acid Res.*, **9**, 1777.
11. Antonarakis,S.E., Orkin,S.H., and Kazazian,H.H. (1982) *Proc. Natl. Acad. Sci. USA*, **79**, 6608.
12. Orkin,S.H., Cheng,T.C., Antonarakis,S.E., and Kazazian,H.H. (1985) *EMBO J.*, **4**, 453.
13. Chang,J.C., and Kan,Y.W. (1979) *Proc. Natl. Acad. Sci. USA*, **76**, 2886.
14. Trecartin,R.F., Liebhaber,S.A., Chang,J.C., Lee,Y.W., and Kan,Y.W. (1981) *J. Clin. Invest.*, **68**, 1012.
15. Weatherall,D.J. and Clegg,J.B. (1975) *B. Biol. Sci.*, **271**, 411.
16. Krawczak,M. and Cooper,D.N. (1991) *Human Genet.*, **86**, 425.
17. Nicholls,R.D., Fischel-Ghodsian,N., and Higgs,D.R. (1987) *Cell*, **49**, 369.
18. Wilkie,A.O.M., Lamb,J., Harris,P.C., Finney,R.D., and Higgs,D.R. (1990) *Nature*, **346**, 868.
19. Wilkie,A.O.M., Buckle,V.J., Harris,P.C., Lamb,J., Barton,N.J., Reeders,S.T.,

Lindenbaum,R., Nicholls,R.D., Barrow,M., Bethlenfalvay,M., Hutz,M.H., Tolmie,J.L., Weatherall,D.J., and Higgs,D.R. (1990) *Am. J. Hum. Genet.*, **46**, 1112.

20. Lamb,J., Wilkie,A.O.M., Harris,P.C., Buckle,V.J., Lindenbaum,R., Barton,N.J., Reeders,S.T., Weatherall,D.J., and Higgs,D.R. (1989) *Lancet*, **ii**, 819.

21. Overhauser,J., Begtsson,U., McMahan,J., Ulm,J.E., Butler,M.G., and Wasmuth,J.J. (1988) *Am. J. Hum. Genet.*, **43**, A92.

22. Wilkie,A.O.M., Zeitlin,H.C., Lindenbaum,R.H., Buckle,V.J., Fischel-Ghodsian,N., Chui,D.H.K., Gardner-Medwin,D., MacGillivray,M.H., Weatherall,D.J., and Higgs,D.R. (1990) *Am. J. Hum. Genet.*, **46**, 1127.

23. Gibbons,R.J., Wilkie,A.O.M., Weatherall,D.J., and Higgs,D.R. (1991) *J. Med. Genet.*, **28**, 729.

24. Lehrman,M.A., Schneider,W.J., Sudhof,T.C., Brown,M.S., Goldstein,J.L., and Russell,D.W. (1985) *Science*, **22**, 140.

25. Lehrman,M.A., Russell,D.W., Goldstein,J.L., and Brown,M.S. (1986) *Proc. Natl. Acad. Sci. USA*, **83**, 3679.

26. Lehrman,M.A., Goldstein,J.L., Russell,D.W., and Brown,M.S. (1987) *Cell*, **48**, 827.

27. Lindenbaum,R.H., Clarke,G., Patel,C., Moncrieff,M., and Hughes,J.T. (1979) *J. Med. Genet.*, **16**, 389.

28. Boyd,Y., Munro,E., Ray,P., Worton,R., Monaco,A.P., Kunkel,L., and Craig,I. (1987) *Clin. Genet.*, **31**, 265.

29. Davies,K.E., Pearson,P.L., Harper,P.S., Murray,J.M., O'Brien,T., Sarfarazi,M., and Williamson,R. (1983) *Nucl. Acid Res.*, **11**, 2303.

30. Kingston,H.M., Sarfarazi,M., Thomas,N.S.T., and Harper,P.S. (1984) *Hum. Genet.*, **67**, 6.

31. Kunkel,L.M., Monaco,A.P., Middlesworth,W., Ochs,H.D., and Latt,S.A. (1985) *Proc. Natl. Acad. Sci. USA*, **82**, 4778.

32. Ray,P.N., Belfall,B., Duff,C., Logan,C., Kean,V., Thompson,M.W., Sylvester,J.E., Gorski,J.L., Schmickel,R.D., and Worton,R.G. (1988) *Nature*, **318**, 672.

33. Monaco,A.P., Neve,R.L., Colletti-Feener,C., Bertelson,C.J., Kurnit,D.M., and Kunkel,L.M. (1986) *Nature*, **323**, 646.

34. Koenig,M., Hoffman,E.P., Bertelson,C.J., Monaco,A.P., Feener,C., and Kunkel,L.M. (1987) *Cell*, **50**, 509.

35. Nudel,U., Robzyk,K., and Yaffe,D. (1988) *Nature*, **331**, 635.

36. Burmeister,M. and Lehrach,H., (1986) *Nature*, **324**, 582.

37. van Ommen,G.-J.B., Verkerk,J.M.H., Hofker,M.H., Monaco,A.P., Kunkel,L.M., Ray,P., Worton,R., Wieringa,B., Bakker,E., and Pearson,P.L. (1986) *Cell*, **47**, 499.

38. Kenwrick,S., Patterson,M.N., Speer,A., Fischbeck,K., and Davies,K. (1987) *Cell*, **48**, 351.

39. Forrest,S.M., Cross,G.S., Thomas,N.S.T., Harper,P.S., Smith,T.J., Read,A.P., Mountford,R.C., Geirsson,R.T., and Davies,K.E. (1987) *Lancet*, **ii**, 1294.

40. Forrest,S.M., Cross,G.S., Speer,A., Gardner-Medwin,D., and Davies,K.E. (1987) *Nature*, **329**, 638.

41. Blondon,L.A.J., Grootscholten,P.M., denDunnen,J.T., *et al.* (1991) *Genomics*, **10**, 631.

42. Love,D.R., England,S.B., Speer,A., Marsden,R.F., Bloomfield,J.F., Roche,A.L., Cross,G.S., Mountford,R.C., Smith,T.J., and Davies,K.E. (1991) *Genomics*, **10**, 57.

43. Abbs,S., Yau,S.C., Clark,S., Mathew,C., and Bobrow,M. (1991) *J. Med. Genet.*, 304.

44. Ried,T., Mahler,V., Vogt,P., Blonden,L., van Ommen,G.-J., Cremer,T., and Cremer,M. (1990) *Hum. Genet.*, **85**, 581.

45. Roberts,R., Bentley,D., Barby,T.F.M., Manner,E., and Bobrow,M. (1990) *Lancet*, **336**, 1523.

46. Hoffman,E.P., Brown,R., and Kunkel,L.M. (1987) *Cell*, **51**, 919.

47. Hoffman,E.P. and Kunkel,L.M. (1989) *Neuron*, **2**, 1019.

48. Monaco,A.P., Bertleson,C.J., Liechti-Gallati,S., Moser,H., and Kunkel,L.M. (1988) *Genomics*, **2**, 90.

49. Love,D.R. and Davies,K.E. (1989) *Mol. Biol. Med.*, **6**, 7.
50. Ryder-Cook,A.S., Sicinski,P., Thomas,K., Davies,K.E., Worton,R.G., Barnard,E.A., Darlinson,M.G., and Barnard,P.J. (1988) *EMBO J.*, **7**, 3017.
51. Lee,C.C., Pearlman,J.A., Chamberlain,J.S., and Caskey,C.T. (1991) *Nature*, **349**, 334.
52. Ascadi,G., Dickson,G., Love,D.R., Jani,A., Walsh,F.S., Gurushinghe,A., Wolff,J.A., and Davies,K.E. (1991) *Nature*, **352**, 815.
53. Valentine,B.A., Cooper,B.J., Cummings,J.F., and le Lahunta,A. (1986) *Acta Neurpath.*, **71**, 301.
54. Gitschier,J., Wood,W.I., Goracka,T.M., Wion,K.L., Chen,E.Y., Eaton,D.H., Vehar,G.A., Capon,D.J., and Lawn,R.M. (1984) *Nature*, **312**, 326.
55. Youssoufian,H., Antonarakis,S.E., Aronis,S., Tsiftis,G., Phillips,D.S., and Kazazian,H.H. (1987) *Proc. Natl. Acad. Sci. USA*, **84**, 3772.
56. Gitschier,J., Wood,W.I., Tuddenham,E.G.D., Shuman,M.A., Goralka,T.M., Chen,E.Y., and Lawn,R.M. (1985) *Nature*, **315**, 427.
57. Kazazian,H.H., Wong,C., Youssoufian,H., Scott,A.F., Phillips,D.G., and Antonarakis,S.E. (1988) *Nature*, **332**, 164.
58. Rommens,J.M., Iannuzzi,M.C., Kerem,B.-S., Drumm,M.L., Melmer,G., Dean,M., Rozmahel,R., Cole,J.L., Kennedy,D., Hidaka,N., Zsiga,M., Buchwald,M., Riordan,J.R., Tsui,L.-C., and Collins,F.S. (1989) *Science*, **245**, 1059.
59. Riordan,J.R., Rommens,J.M., Kerem,B.-S., Alon,N., Rozmahel,R., Grzelczak,Z., Zielenski,J., Lok,S., Plavsic,N., Chou,J.-L., Drumm,M.L., Iannuzzi,M.C., Collins,F.S., and Tsui,L.-C. (1989) *Science*, **245**, 1066.
60. Higgins,C.F., Hyde,S.C., and Gill,D.R. (1991) *Trends Biochem.*, in preparation.
61. Tsui,L.C., and Estivill,S. (1991) In *Genome Analysis, Volume III*, K.E.Davies and S.M.Tilghman. Cold Spring Harbor Laboratory, New York.
62. Tsui,L.-C. (1992) Current Biology in Current Opinions in Genetics and Development, in press.
63. Higgins,C.F. and Hyde,S.C. (1991) *Nature*, **352**, 194.
64. Wallace,M., Marchuk,D., Andersen,L., Lechter,R., Odeh,H., Saulino,A., Fountain,J., Brereton,A., Nicholson,J., Mitchell,A., Brownstein,B., and Collins,F.C. (1990) *Science*, **249**, 181.
65. Viskochill,D., Buchberg,A.M., Xu,G., Cawthorn,R.M., Stevens,J., Wolff,R.K., Culver,M., Carey,J., Copeland,N.G., Jenkins,N.A., White,R., and O'Connell,P. (1990) *Cell*, **62**, 187.
66. Cawthorn,R.M., Weiss,R., Xu,G., Viskochil,D., Culver,M., Stevens,J., Robertson,M., Dunn,D., Gesteland,R., O'Connell,P., and White,R. (1990) *Cell*, **62**, 193.
67. Ballester,R., Marchuk,D., Boguski, M., Saulino,A., Letcher,R., Wigler,M., and Collins,F. (1990) *Cell*, **63**, 851.
68. Hall,J.G. (1990) *Am. J. Hum. Genet.*, **46**, 857.
69. Moore,T. and Haig,D. (1991) *Trends Genet.*, **7**, 45.
70. Cattenach,B.M. (1991) In *Genome analysis, Volume II*, (ed. K.E.Davies and S.M.Tilghman). Cold Spring Harbor Laboratory, New York.
71. Sapienza,C. (1990) *Scientific American*, October.
72. Nicholls,R.D., Fischel-Ghodsian,N., and Higgs,D.R. (1987) *Cell*, **49**, 369.
73. Scrable,D., Cavanee,W., Ghavimi,F., Lovell,M,Morgan,K., and Sapienza,C. (1989) *Proc. Nat. Acad. Sci. USA*, 7480.
74. Fryns,J.-P. (1990) in *The fragile X syndrome*, (ed. K.Davies), pp. 1–39. Oxford University Press.
75. Sherman,S.L., MortonN.E., Jacobs,P.A., and Turner,G. (1984) *Ann. Hum. Genet.*, **48**, 21.
76. Sherman,S.L., Jacobs,P.A., Morton,N.E., Froster-Iskenius,U., Howard-Peebles,P.N., Nielsen,K.B., Partington,M.W., Sutherland,G.R., Turner,G., and Watson,M. (1985) *Hum. Genet.*, **69**, 289.

77. Lubs,H.A. (1969) *Am. J. Hum. Genet.*, **21**, 231.
78. Sutherland,G.R. (1977) *Science*, **197**, 256.
79. Pembrey,M.E., Winter,R.N., and Davies,K.E. (1985) *Am. J. Med. Genet.*, **21**, 709.
80. Nussbaum,R.L., Airhart,S.D., and Ledbetter,D.H. (1986) *Am. J. Med. Genet.*, **23**, 715.
81. Laird,C., Jaffe,E., Karpen,G., Lamb,M., and Nelson,R. (1987) *Trends Genet.*, **3**, 274.
82. Vincent,A., Heitz,D., Petit,C., Kretz,C., Oberle,I., and Mandel,J.-L. (1991) *Nature*, **329**, 624.
83. Bell,M.V., Hirst,M.C., Nakahori,Y., MacKinnon,R.C., Roche,A., Flint,T.J., Jacobs,P.A., Tommerup,N., Tranebjaerg,L., Froster-Iskenius,U., Kerr,B., Turner,G., Lindenbaum,R.H., Winter,R., Pembrey,M., Thibodeau,S., and Davies,K.E. (1991) *Cell*, **64**, 861.
84. Yu,S., Pritchard,M., Kremer,E., Lynch,M., Nancarrow,J., Baker,E., Holman,K., Mulley,J.C., Warren,S.T., Schlessinger,D., Sutherland,G.R., and Richards,R.I. (1991) *Science*, **252**, 1179.
85. Oberle,I., Rousseau,F., Heitz,D., Kretz,C., Devys,D., Hanauer,A., Boue,J., Bertheas,M., and Mandel,J.L. *Science*, **252**, 1097–102.
86. Verkerk,A.J.M.H., Pieretti,M., Sutcliffe,J.S., Fu,Y.H., Kuhl,D.P.A., Pizzuti,A., Reiner,O., Richards,S., Victoria,M.F., Zhang,F., Eussen,B.E., van Ommen,G.-J.B., Blonden,L.A.J., Riggins,G.J., Chastain,J.L., Kunst,C.B., Galjaard,H., Caskey,C.T., Nelson,D.L., Oostra,B.A., and Warren,S.T. (1991) *Cell*, **65**, 905.
87. Nakahori,Y., Knight,S.J.L., Holland,J., Schwartz,C., Roche,A., Tarleton,J., Wong,S., Flint,T.J., Froster-Iskenius,U., Bentley,D., Davies,K.E., and Hirst,M.C. (1991) *Nucl. Acid Res.*, **19**, 4355–9.
88. Davies,K.E. (1992) *Nature*, **356**, 15.
89. Todd,J.A., Aitman,T.J., Cornall,R.J., Ghosh,S., Hall,J.R.S., Hearne,C.M., Knight,A.M., Love,J.M., McAleer,,M.A., Prins,J.-B., Rodrigues,N., Lathrop,M., Pressey,A., Delarato,N.H., Petersen,L.B., and Wicker,L.S. (1991) *Nature*, **351**, 542.
90. Partridge,T.A., Morgan,J.E., Coulton,G.R., Hoffman,E.P., and Kunkel,L.M. (1989) *Nature*, **337**, 176.
91. Rinchik,E.M. and Russell,L.B. (1991) In *Genome analysis, Volume I,* (ed. K.E.Davies and S.M.Tilghman), Cold Spring Harbor Laboratory, New York.
92. Rudnicki,M.A. and Jaenisch,R. (1991) In *Genome analysis, Volume 11,* (ed. K.E.Davies and S.M.Tilghman). Cold Spring Harbor Laboratory, New York.
93. Jaenisch,R. (1988) *Science*, **240**, 1468.
94. Stacey,A., Bateman,J., Choi,T., Mascara,T., Cole,W., and Jaenisch,R. (1990) *Nature*, **332**, 131.
95. Evans,M. (1991) In *Genome analysis, Volume II,* (ed. K.E.Davies and S.M.Tilghman). Cold Spring Harbor Laboratory, New York.
96. Capecchi,M.R., (1991) *Trends Genet.*, **5**, 70–6.
97. Friedman,T. (ed.) (1991) In *Therapy for genetic disease*, p.107. Oxford University Press.
98. Vega,M.A. (1991) *Human Genet.*, **87**, 245.
99. Rosenberg,S.A., Aebesold,P., Cornetta,K., Kasid,A., Morgan,R., Moen,R., Karson,E.M., Lotze,M.T., Yang,J.C., Topalian,S.L., Merino,M.J., Culver,K., Miller,A.D., Blaese,R.M., and Anderson,W.F. (1990) *New Engl. J. Med.*, **323**, 570.

Glossary

Alleles: alternative forms of a gene which segregate at meiosis, e.g. the A, B, and O forms of the ABO blood group gene.

***Alu* sequence:** sequence of DNA recognized by the restriction enzyme *Alu*I which is repeated about 300 000 times in the human genome.

Amber mutations: mutations that can be suppressed. An amber mutation results in the creation of a UAG stop codon in the mRNA, indicating termination of translation. The mutation can be suppressed in certain strains of *E.coli* possessing a tRNA with the CUA anticodon which inserts an amino acid at UAG site permitting continued translation.

Autosome: any chromosome except the X or Y chromosomes.

Bayesian statistics: a method for combining several independent likelihoods to get an overall probability.

Chromosome walking: sequential isolation of clones which contain overlapping DNA sequences so as to move along the chromosome.

Consanguineous: all humans are ultimately related, but a marriage is described as consanguineous if the couple are particularly closely related, e.g. cousins.

Consanguineous pedigree: a pedigree in which related individuals have married and produced offspring.

Diploid: a cell containing two genomes, like most normal human body cells.

Exon: region of the gene containing the coding sequence which is transcribed into the precursor RNA and remains after RNA processing.

Genetic drift: in evolution, the tendency for gene frequencies in small populations to vary randomly between generations.

Haplotype: a particular combination of alleles at a set of closely linked loci which tend to be transmitted as a block.

Heterokaryon: cell line resulting from fusion of nuclei from two different species.

Homologous: of chromosomes, the two of a pair, e.g. the two chromosome 1's.

Horizontal inheritance: a pedigree pattern where the affected people are all in the same generation (usually sibs).

Housekeeping genes: genes expressed by essentially all cells constitutively, e.g., genes coding for many metabolic enzymes.

Intron (intervening sequence): region of the gene which is transcribed but is spliced out of the precursor RNA in the production of mature mRNA.

Linkage: loci are linked if they lie sufficiently close together on a chromosome that they tend to segregate together.

Locus: the location of a gene on a chromosome; the *ABO* locus can be occupied by the *A, B,* or *O* alleles.

Lod score: the logarithm of the odds that two loci are linked.

Lyonization: X chromosome inactivation in a cell having more than one X chromosome.

Marker: a polymorphic genetic character used to follow the transmission of a chromosomal segment in a family or population.

Oligoprobe: a short DNA probe whose hybridization is sensitive to a single-base mismatch.

Open reading frame (*orf*): DNA sequence which corresponds to potential coding information of a protein.

Penetrance: the probability that a disease genotype will cause the disease.

Phase: in a person doubly heterozygous for linked markers (*AaBb*), the two possible phases are *AB/ab* and *Ab/aB*.

Phenotype: the observable characteristic of a person (affected, type 2, high-responder, etc.).

Polymorphic: having more than one common form within a population.

Proband: the person through whom the pedigree was ascertained; often the person requiring a risk estimate.

Recombinant: in linkage analysis, having a new combination of linked markers on a chromosome because of crossing-over during a parental meiosis.

RFLP: polymorphic variants in the size of fragments produced by digesting DNA with a restriction enzyme.

Sibs: brothers or sisters.

Splicing: the removal of the introns and joining of exons in RNA processing to give mature mRNA.

Translocation: an abnormal chromosomal arrangement in which parts of two different chromosomes are joined.

Trait: any genetic character, whether a disease or a normal variation like blood group, etc.

Vertical inheritance: a pedigree pattern where people in several successive generations are affected.

Zoo blot: Southern blot of genomic DNA from several different eukaryotic species.

Glossary of genetic diseases

These very brief descriptions are for orientation only; refer to a medical textbook for the full range of manifestations.

Achondroplasia: short-limbed dwarfism; normal size body and head. Normal intelligence. The homozygous state is lethal.

α-thalassaemia: reduced or absent α-globin synthesis. Complete absence of α chains is lethal.

Becker muscular dystrophy: as Duchenne muscular dystrophy but slower. Spectrum of severity from near Duchenne to near normal.

β-thalassaemia: absence or deficiency of β-globin synthesis. Persistent anaemia requiring repeated transfusions, which lead to problems with iron overload.

Chronic granulomatous disease: white cells are unable to kill bacteria, leading to recurrent infections and death in childhood.

Cirrhosis: progressive liver failure leading to jaundice; can be fatal. Has many different causes.

Cystic fibrosis: all mucous secretions are too viscid, resulting in recurrent lung congestion and infections, and malabsorption of food from pancreatic insufficiency. Fatal by early adulthood.

Down's syndrome: mental retardation, characteristic build and face, due to an extra copy of chromosome 21.

Duchenne muscular dystrophy: slow to walk, difficulty running and climbing stairs. Progressive muscle weakness, wheelchair by age 10, death by age 20. Some but not all mentally retarded.

Emphysema: Shortness of breath due to breakdown of alveolar structure of lungs. Has many different causes.

Familial hypercholesterolaemia: heterozygotes have raised cholesterol levels on normal diet, high risk of heart disease in middle life. Homozygotes likely to die of heart attacks in twenties.

Focal dermal hypoplasia: rare disease affecting females only; abnormalities of hands, feet, teeth; patchy skin defects and pigmentation; papillomas (like warts) on mucous membranes.

Haemophilia A: blood will not clot because of low Factor VIII activity. Can be crippling because of recurrent bleeding into joints. Severity quite variable.

Huntington's disease: progressively worse uncontrolled movements and dementia. Symptoms typically start in the forties, but age of onset is very variable. Death 12 – 15 years later.

Hypophosphataemic rickets: kidneys fail to retain phosphate, leading to rickets unresponsive to normal doses of vitamin D. Milder and more variable in females than in males.

Incontinentia pigmentii: rare disease affecting females only. Skin has pigmented streaks and patches; sometimes mental retardation, fits, and skeletal abnormalities.

McLeod phenotype: abnormal shape of red cells (acanthocytosis) with haemolytic anaemia and absence of red cell Xk antigen.

Prader – Willi syndrome: mental retardation and hypotonia (floppiness); later uncontrolled eating and gross obesity. Males have underdeveloped sexual organs.

Retinitis pigmentosa: progressive loss of vision, starting as night blindness. Many different genetic forms.

Retinoblastoma: childhood tumour of the retina, fatal if eye not removed. Unilateral or bilateral.

Rett syndrome: rare and mysterious disease affecting only females. Progressive mental changes in childhood leading to autistic state with no communication and repetitive movements.

Sickle cell disease: red blood cells collapse and clump together, causing anaemia and blocking small vessels.

Tay – Sachs disease: normal in infancy but then progressive neurological degeneration leading inevitably to death in childhood.

von Recklinghausen neurofibromatosis: very variable disease. At worst, constant growth of grossly disfiguring nodules all over skin, other organs also affected. At best, a few pigmented skin spots.

Wilm's tumour: childhood tumour of kidney; usually complete cure if removed soon enough.

Xg blood group: a minor antigen not important in transfusion. Presence or absence of Xg(a) antigen on red cells.

Index